DIGITAL
TRANSFORMATION

EVOLVING A DIGITALLY ENABLED NIGERIAN PUBLIC SERVICE

JACOBS EDO
FOREWORD: AXEL UHL & ALEXANDER SAMARIN

me@jacobsedo.com
www.digitaltransformation.com.ng

Book Cover Design by Dejan Popov
Book Layout & Design by JP Kusmin
E-Book Layout & Design by Vassilaco
Set in 11.5 point Utopia

First Printed, September 2016
ISBN-13: 978-0-9977624-4-0 (Kindle)
ISBN-13: 978-0-9977624-3-3 (Hardback)
ISBN-13: 978-0997762440 (Paperback)
ISBN-13: 978-09977624-0-2 (E-book)

Disclaimer

The links or websites in this book are provided for informational purposes only and do not constitute an endorsement of any products or services provided by these websites and that the links are subject to change, expire, or be redirected without any notice.

This book was written in a personal capacity. Opinions and submissions contained herein are the author's and do not represent the views of the OPEC Fund for International Development (OFID). Responsibility for errors and omissions equally remains with the author.

Dedication

This work is dedicated to my dear wife, Matilda,
my daughter, Judith Eseosa and my sons, Joseph
Eghosasere, Jeffrey Efeosa and my mentor
Dr Deyaa Lutfi Alkhateeb for their unending love,
understanding and support.

Acknowledgments

This book originated from a conversation with Saad Ahmed about moving Nigeria forward. Throughout the process of writing this book, many individuals and friends have taken time out to assist me in one way or the other. I would like to give special thanks to Esema Aguse, Joannes Vlachos, Iheanacho Nzurum, John Offikwu, Olanrewaju Fagbodun, Richard Egobi and Paul Oluwabunmi for actively participating in the feedback and contributions for this book.

I also thank Samuel Ifeagwu – my amiable senior colleague and proofreader – and Dejan Popov for producing an excellent book design. So much hard work has gone into the Digital Transformation arena that it has become one of the most robust and widely used terminologies by tomorrow's leaders.

I remain grateful to David Chin, Jonathan Dimson, Andrew Goodman and Ian Gleeson of McKinsey and company on their work: Worldclass Government: Transforming the UK Public Sector in an Era of Austerity: Five lessons from around the world, which formed a solid basis for this book's research commencement.

I would like to offer extra special thanks to the rest of the individuals in this section. Without your insights or contributions, portions of this book may not have been possible – Amieyeofori Solomon, Kenneth Omeruo, Akanimo Udoh, Wayne Haw, Larry Chris Bates, Anthony Oyovwe, Michael Ebiye, and Ngozi Fejokwu.

Thanks to Rex Edo - without you, this book would never find its way to the Web nor the Internet. Last and not least, I beg forgiveness of all those who have been with me over the course of the past years and whose names I may have failed to mention.

"It is not the strongest of the species that survive, nor the most intelligent, but the one most responsive to change" — Charles Darwin

Foreword

Since meeting Jacobs in 2014 at a Global Business Transformation event in Potsdam, Germany, his passion for applying his hard-earned digital business transformation knowledge to improve the lives of his fellow citizens of Nigeria has been evident. This book is an excellent contribution to his country, which could benefit immensely from the guidance and research that Jacobs presents to those who are ready to learn and make a positive difference in Nigeria.

The old saying that "Rome wasn't built in a day" refers to a real city that took time to build. Similarly, it has taken time to build brick and mortar companies and countries, because they have traditionally relied upon expensive physical assets to grow.

After thousands of years of people, business and nations thriving (or not) according to their physical assets, the ability to innovate, digitise, and transform has finally changed the game. It has levelled the playing field allowing underdogs to disrupt the status quo, without the physical constraints of the past.

In this book, Jacobs highlights Nigeria's opportunity to embrace Digitalisation, and equipped with its strengths, to take advantage of the level playing field that now stands before it. But a sense of urgency is vital before the digital economy becomes more of a threat than the current opportunity it presents Nigeria with. This feeling of urgency needs to be driven by President Muhammadu Buhari and adopted by his most senior team.

Nigeria can take inspiration from countries like Estonia, which Jacobs describes in the book as being one of the world's most digital societies, thanks to a government that were quick to embrace the digital economy. Estonia is small and without the physical assets of many other nations, and yet it has demonstrated how this is no longer a reason for countries to be held back by the physical constraints of the past.

Nigeria has an opportunity to mature its e-government development index, and in doing so, provide Nigerian start-ups and existing companies with the foundations upon which they can build new digital offerings that can be easily purchased throughout the world. Because digital offerings have no physical borders to cross, and almost no storage and transport costs to absorb.

The digital economy presents Nigeria with an opportunity to open up new channels that will enable revenues to pour in from around the world for Nigeria's potential digital offerings. In turn, this can help the country flourish in a new era. But this potential needs to be encouraged and supported by the Nigerian government in the way of infrastructure, funding, and education for the next generation of digital entrepreneurs to thrive in the global digital economy.

It will be Nigerian entrepreneurs and business leaders who create the country's digital offerings for the world to buy, but this will be difficult without the right support from government.

Learn and become inspired by this book - then act with a sense of urgency to help equip Nigeria's people for the global digital opportunity that awaits them.

Rob Llewellyn is the co-author of a new digital disruption book to be published by Springer, a keynote speaker at digital transformation conferences, and has been awarded Global Business Transformation Master status by SAP as a trusted C-suite advisor. He is also the creator of the THRIVE Digital Business Transformation Framework and its accompanying premium online video training for executives.

August 2016.

*"Change is the law of life, and those who look only to
the past or present are certain to miss the future"*
— *John F. Kennedy*

Companies across the globe and industries keep transforming their businesses to stay ahead of the game. Conversely, if a company does not see the need to change or does not make the right changes, it soon ceases to exist. As a result of economic and technical potentials, enabled by phenomena such as digital media, big data, mobile and cloud technologies, the need for change over the next twenty years will be bigger than in all of the last 300 years combined.

Change is no longer linear; it is exponential – and we are still at an early stage. It impacts companies' strategies, structure, culture, products, services, processes, and people.

Overall, digital business is one of the fastest growing industries. Online sales, for example, increase by 40 per cent every year. But not all companies in all countries benefit in the same way. Particularly the most advanced economies such as the United Kingdom, the United States of America, South Korea, and Japan take advantage of this trend, whereas less developed countries are unable to leverage the potential appropriately.

Digital technologies also lay the foundation for a new industrial revolution. The German concept 'Industrie 4.0', also known as 'Smart Industry', integrates manufacturing with state-of-the-art information and communication technology. This smart approach makes it possible to deliver products that are tailored to meet individual customer requirements – at low cost and high quality.

'Industrie 4.0' and the Internet of Things have become real game changers to many industrialised countries and open exciting new opportunities for countries worldwide. This great potential is why it is so important to put all our energy into digitalisation. How well a country manages to make progress in this area will decide its future – no more and no less.

The key question we have to ask ourselves is this: What makes some countries succeed in their digital transformation efforts where others fail? Before we can answer this question, we have to understand that Digitalisation is not just about implementing new technology. It is about stakeholder management, intensive change management, and effective communication with all interest groups.

With his book about digitalisation in Nigeria, Jacobs Edo provides us with answers. His aim is to encourage stakeholders in public institutions to focus their efforts on creating more and better digital services. Far from just describing the benefits of digital transformation, this book also discusses important findings from a variety of successful digital governance initiatives across the globe. It also offers suggestions on how to increase the long-term digital readiness of Nigeria. In doing so, Jacobs Edo makes a valuable contribution towards ensuring a successful future for Nigeria.

Professor Dr Axel Uhl is the deputy head of the Institute for Business Information Technology at the Zurich University of Applied Sciences. His main areas of research are Digital Transformation, Business Transformation, Sustainability and Leadership. Before working at the University, he had been in Senior Management roles at Allianz Insurance AG, DaimlerChrysler AG, KPMG, Novartis and SAP SE.

August 2016.

Table of Contents

Table of Figures

Tables

"Disruptors don't set out to beat you at
your own game - they change the rules."
— *Kai Rieme*

Preface

Digitalization is here to stay.

Adapting principles from the old regime that I consider the analogue era, over time will remain essential to corporate and institutional success – whether in the public or private sector. According to a recent McKinsey report [1], the average corporate lifespan has been falling for more than half a century. Standard and Poor's (S&P) data show that companies' life-spans contracted in the last century. From 1958 when it was an average of 61 years, it fell to 25 years in 1980 and just about 18 years in 2011, by reasons of digital transformation.

Digitalization is placing unprecedented pressure on organisations and government institutions to evolve. At the present rate, 75% of the S&P 500 incumbents would be gone by 2027 [2]. That means managing our collective transition to a digitally-driven business model is vital and crucial to our quest for a prosperous nation. And since digital touches so many parts of our lives, society and development, any significant reform program requires coordination of people, processes, and technologies. This transition toward a successful value delivery needs new skills, a holistic vision on growth trends and challenges, and a seamless collaboration between technology and governance.

This book therefore defines, justifies and outlines the urgent steps needed for digital transformation in the Nigerian public service. At the heart of this clarion call is the on-going civil service reform, a process which has yet to formulate and implement a comprehensive as well as an integrated strategy to digitise public infrastructure and services – a matter which should be of national urgency.

The process of digital transformation encompasses, by definition, two concepts: Digitalisation and Transformation. While digitalisation is about making Information and Communication Technologies (ICT) integral to the government's function whether federal, state, or local government, business process transformation applies to addressing the weaknesses that continually challenge public service operations. A digital transformation of the Nigerian public service would enable it to embrace the much-needed change; including digital technologies and other innovative approaches to improve service delivery and management as well as its working culture, and redefine the value systems in the public service for a worthwhile and rewarding outcome.

The Nigerian public service has real strength and potential. It exists to implement the policies of the government of the day, regardless of that government's political colouration. Its position and expected political impartiality enable exceptionally rapid transitions between governments. The majority of Civil Servants are dedicated, hard-working and have a deep-seated public service ethos but change has come to Nigeria as per recent developments.

The Nigerian public wants quality services to be delivered faster, better, greener, cheaper, more integrated and more ecologically friendly. Nigeria's dependence on ever-fluctuating oil prices, which are today near their lowest in recent economic history, entails that these improvements must be delivered at a lower cost than ever before. This means that the drive for greater efficiency must be relentless, and productivity must continue to improve. Public sector productivity has often been static (and sometimes regressive) whilst the private sector has grown year after year. The need for efficiency and economic sustainability implies that government work, wherever possible, must become digitally enabled and integrated within its numerous Ministries, Departments and Agencies (MDAs).

Nowadays, the best service institutions deliver online everything that can be offered online. The concept of "digital first" in the private sector cuts operational costs dramatically and allow access to information, knowledge and services at times and in ways that are user or people-centred. This is at variance with the provider-centred model, the antithesis to previously dominant models of public service delivery as inherited from the colonial masters.

The government's ICT capacity has lagged far behind that of the private sector for too long. Although fledgeling and successful stand-alone heterogeneous technology efforts can be found in many government MDAs, they are insufficient and piecemeal. Therefore, the pace of change needed to catch up with contemporary international standards places significant demands on the civil service, its administration and its staff. Public servants will require better skills, better tools and a mindset that is firmly citizen or user-focused to engage the changing global economic climate successfully.

While the profound uncertainty surrounding the development and adoption of emerging technologies means that we do not yet know how the transformations driven by this industrial revolution will unfold, their complexity and interconnectedness across sectors imply that all stakeholders of global society- governments, business, academia, and civil society – have a responsibility to work together to better understand the emerging trends. Shared understanding is particularly critical if we are to shape a collective future that reflects common objectives and values. We must have a comprehensive and globally shared view of how technology is changing our lives and those of future generations, and how it is reshaping the economic, social, cultural and human context in which we live. The changes are so profound that, from the perspective of human history, there has never been a time of greater promise or potential peril. My concern, however, is that decision-makers are too often caught in traditional, linear (and non-disruptive) thinking or too absorbed by immediate concerns to think strategically about the forces of disruption and innovation shaping our future [3].

For Nigeria to embrace the Sustainable Development Goals, its Vision 20:20:20, the hopes and aspirations of its young population and the public servants, the government will need to embark on a general public sector digital transformation programme on a scale unprecedented on the African continent. This work outlines the proposed goals and objectives of this process; it also sets forth the following public service transformational vision:

> A fully functional and agile public service administration resolutely oriented towards the Nigerian citizen, with adequately resourced core services providing a nurturing and rewarding working environment for public servants and high-quality services to citizens.

Also, the under listed approaches, taken together, could constitute the core agenda for the anticipated change in an era of continued public service decline:

1. Optimise the overall federal, state and local governments' structure, scale and operating business model for digital enablement;
2. Radically redesign public services to improve quality of service and cost efficiency;
3. Restructure the government's approach to public service delivery incorporating a common shared service model and framework;
4. Strengthen functional/technical leadership and capabilities across government services to support efficient delivery; and
5. Develop the vision, accountability and capacity needed to drive a public sector digital transformation effort.

Around the world, governments are embracing digital transformation through digital service enablement. Government digital service is becoming the norm and is central to technology adoption in the public service globally. It is also true that wealthy nations are exploring the possibility of using blockchain technology, an idea that underpins the bitcoin crypto-currency to increase efficiency in the delivery of high quality and trust-based services. A blockchain works as a decentralised ledger that is verified and shared by a network of computers, and can be used to record data as well as to secure and validate transactions. Banks and other financial institutions are increasingly investing in blockchain technology, knowing it would cut their costs and make their operations faster and more transparent.

The government of Nigeria is not the first to face the need for fiscal consolidation or improvement in public service delivery. Sweden, Denmark, Australia and Israel all recovered from significant challenges and budget deficits in the 1990s and 2000s. Similarly, the United States of America, Germany and Kenya have taken important steps to improve service delivery and management on tight budgets [4].

According to the Greeks, "A society grows great when old men plant trees whose shade they know they shall never sit in." Nigerian government leaders are therefore invited to consider seriously digitising their public services as detailed in this book.

Jacobs Edo,
Vienna, Austria.
August 2016.

"If a better society is to be built, one that is more just and more loving, one that provides greater creative opportunity for its people, then the most open course is to raise both capacity to serve and the vary performance as servant of existing major institutions by new regenerative forces operating within them"
— Greenleaf, 1970

INTRODUCTION

*"I cannot understand why people are frightened of
new ideas. I'm frightened of the old ones."*
— *John Cage*

Chapter 1

·············•◆•·············

INTRODUCTION

Since 1945 various public sector reforms have been attempted; however, the Nigerian public service remains riddled with inefficiency, lack of financial accountability, poor human resource management and a disregard for its clients – the citizens. Some of the critical issues facing the public service are a lack of prioritisation of citizens' needs and the misallocation of human and financial resources that follow these misplaced priorities.

For the country to embrace the Sustainable Development Goals, its own Vision 20:20:20, the hopes and aspirations of its citizens and its public servants, the public service must become efficient, more strategic and people-centred. Its new organisational model must be flexible and malleable to government programmes and the needs of the people.

This book outlines the goals and objectives of this digital transformation process; it also sets forth the following vision:

> A fully functional and agile public service administration resolutely oriented towards the Nigerian citizen, with adequately resourced core services providing a nurturing and rewarding working environment for public servants and high-quality services to citizens.

The Nigerian government is presently running a strategy for reforming its public services. Digital transformation can make tangible contributions to each of the goals listed in the National Strategy for Public Sector Reform (NSPSR), as well as to the fulfilment of its mission statement. This book's implementation approach rests on eight principal activities each led by operational objectives and key performance indicators.

Despite a chequered history of government digitalisation activities, a development dating over 15 years, Nigeria only has a digital government development index of 0.2929 and ranks at 141 out of 193 countries surveyed in 2014 by the United Nations.

Country	Level of Income	EGDI	2014 Rank	2012 Rank	Change in Rank
High EGDI					
Tunisia	Upper Middle	0.5390	75	103	↑28
Mauritius	Upper Middle	0.5338	76	93	↑17
Egypt	Lower Middle	0.5129	80	107	↑27
Seychelles	Upper Middle	0.5113	81	84	↑ 3
Marocco	Lower Middle	0.5060	82	120	↑38
Middle EGDI					
South Africa	Upper Middle	0.4869	93	101	↑ 8
Botswana	Upper Middle	0.4198	112	121	↑ 9
Namibia	Upper Middle	0.3880	117	123	↑ 6
Kenya	Low	0.3805	119	119	-
Libya	Upper Middle	0.3753	121	191	↑70
Ghana	Lower Middle	0.3589	123	145	↑22
Rwanda	Low	0.3589	125	140	↑15
Zimbabwe	Low	0.3585	126	133	↑ 7
Cape Verde	Lower Middle	0.3551	127	118	↓ 9
Gabon	Upper Middle	0.3294	131	129	↓ 2
Algeria	Upper Middle	0.3106	136	132	↓ 4
Swaziland	Lower Middle	0.3056	138	144	↑ 6
Angola	Upper Middle	0.2970	140	142	↑ 2
Nigeria	Lower Middle	0.2929	141	162	↑21
Cameroon	Lower Middle	0.2782	144	147	↑ 3
Regional Average		**0.2661**			
World Average		**0.4712**			

Figure 1: World E-Government Ranking for Africa [5]

These unimpressive numbers can be attributed to a lack of understanding of the importance of digital technologies and their disruptive effect on the government as well as a lack of leadership in the digital transformation arena.

Digital public services refer to "a public service delivered electronically over the Internet that offers a simple and accessible way for citizens to transact with the government" [6]. However, digital transformation encompasses far more, including the transformation of all aspects of government operations and support services. A digital government will cover the full range of digitalisation - from the core digitalisation of public services to digital infrastructure, governance and processes, including both front and back-office transformation. Digitalisation will allow the government to operate with greater transparency and efficiency. It will also improve access to essential services and information. Above all, a digital government will be user or citizen centred.

The renewed call for digital transformation will face many challenges, including from the civil service environment itself and from the diverse demographic that is its client base, as well as funding and leadership challenges. For a nation that is composed mostly of young people, the concept that the public sector can be effective, noble, helpful and that it can have the interests of the people at heart is an attractive preposition. Improving civil service culture will be a difficult undertaking, and one that will require a committed leadership, one that can visualise the potential of digital transformation.

AIM OF THE BOOK

"This year's theme at the meeting in Davos was about "mastering the 4th industrial revolution" and what that really means is trying to make sense of the implication of the digital transformation the world is going through right now, and making sure we can collectively ensure that it serves humanity and doesn't threaten it," - Vanessa Moungar, Community Lead, Regional Strategies Africa for WEF [7].

This book is targeted at high-level decision makers particularly in Nigeria, emerging markets and other developing countries. It aims to facilitate and stimulate wider discussions and the pursuit of digital transformation objectives as they relate to economic and social development especially harnessing the power of Information and Communication Technologies (ICT).

The book exposes specific areas of the Nigerian public service, government agencies, education, health and social services where new waves of digital transformation could be gainfully applied to combat poverty, eliminate corruption and elevate the standard of living of all Nigerians. Also, the book reveals how digital transformation can be employed to improve the security of life and properties in Nigeria.

It also encourages private sector entrepreneurs and SMEs to capitalise on potential opportunities derived from digital transformation thereby creating employment for the teeming youth population of Nigeria, and indeed the youth population of the African continent. This, the book argues, is achieved by sensitising the government into legislating frameworks and making policies that enable the realisation of benefits of digital transformation in different sectors of the Nigerian economy.

And finally, an important aim of the book is to expose areas where digital transformation can help in the fight against poverty. The book shows how this could be achieved through implementation of paper-less and faceless transactions within government agencies and private sectors linked to an already existing centralised bank of personal data at Government disposal. Personal data on National Identity, Bank Verification Number and driver's license can be harnessed to effect auditable transaction trails backed by policy-driven service level agreements between government, service providers and the general public.

According to Marc Saxer, "the perplexing paradox about digital trans-formation is that while the way we work, live and think is about to change radically, the nature of the political class remains unaffected. While the organisations, institutions and policies built for the indus-trial age are about to vanish, the progressive idea of the political class as the agonistic struggle between political projects is more up to date than ever. What is needed, therefore, is nothing less than a new progressive project for the digital age. It is high time for progressives of all tribes to rediscover political economy and practical utopia, join forces, and prepare for the struggle over the shape of the digital trans-formation. This struggle will define the world of tomorrow." [8]

In the end, the only way to overcome a transformation conflict is concluding a social and economic contract based on inclusiveness for all especially the poor amongst us.

WHY THIS BOOK

A digital transformation in Nigeria is the Big Exit, the ajar door, from the mass youth unemployment. One, it is directly relates to an ICT revolution and penetration providing for a mass employment of labour, enablement of potentials and realisation of opportunities. Even though the pursuit of this book is the public sector, how it can influence the private sector has been severally alluded to or directly encouraged.

A practical example of how this is possible is the everyday reality of many young Nigerians. Take, for instance, the outlier **Yvonne Anoruo**.

Yvonne's primary tool for her business is the Internet. She's neither an engineer nor trained in any of the sciences. In fact, she turned her Maths answer booklet blank in an examination. She's an English graduate of the University of Lagos. She also isn't following her parent's footstep in driving business. Neither has she gone to a business school. But she's at the head of a team employing ICT, market skills and a combo of authoritative and affiliative leadership styles to deliver on market objectives. She's self-evolved, defining a possibility for many a young Nigerian.

Anybody, like her, not trained in STEM (Science, Engineering, Technology and Maths) would have been an outlier but as she is demonstrating, at www.efiritin.com, the internet, framed in digitalization, is not only an enabler – it's also a leveller. There are no outliers -- just a cosmos of opportunities for an English graduate as well as for an ICT folk hero.

Yvonne belongs to a new generation of Nigerians who, unsupported by the government and the systems, are creating value by transformative entrepreneurship and smart initiatives.

But like other Nigerians, the challenge of relating to an un-digitized public service, using slow internet connectivity and opportunities missed in low digitised society handicaps her. Her team's performance can be higher. Pursued to a logical conclusion, productivity drops for everything and everybody associated with the internet in Nigeria. That's macroeconomic. Output drops. GDP falls. Jobs are lost.

If national development for a country as Nigeria is about lifting populations out of poverty, a digital transformation stands as a ranking platform to deliver on this goal. Government's business is simply to create exit so that the economically condemned can work their way out. How does a government create exit? A practical, simple way is heading into public service digitalization.

This direction would be one movement that signifies progress, one drive that measures its weight in gold and one policy that would re-enforce its sustainability and orchestrate a domino or influence across other economic sectors and service delivery.

HOW THIS BOOK IS ORGANIZED

The book is structured along three major sections. Section I is structure to provide foundational information around the book's context. Section II is designed to present strategy and plan in step form for digital transformation project design and planning. While section III covers some designated implementation approaches complete with expected outcomes and indicators.

The outline of the book are as follows:

Content

Foreword
- Rob Llewellyn
- Prof Dr Axel Uhl

Preface

Chapter 1: Introduction

- Aim of the Book
- Why this Book
- How the Book is organised

Section I: Background and Service Review

Chapter 2: Nigeria: A Brief

This chapter sets up the challenge facing the Nigerian Civil Service by beginning with a presentation of the place. The chapter then describes the political situation and the economics that impact the governance, infrastructure, and education of the country. Digital government is introduced as an opportunity for growth and change. The chapter ends with a strong expression of the need for change, stopping short of outlining each of the challenges.

- Nigeria Politics and Economy
- Digital in Nigeria
- The Civil and Public Service Today
- The History of Civil Service Reform
- The Need for Change in the Nigerian Civil Service

Chapter 3: Digital Transformation Lessons

This chapter outlines the digital transformation process, principles, and experiences from around the globe. It describes approaches and impact that can be used to both guide and evaluate planning and implementation in multiple countries. The chapter ends with a review of the digital transformation principles in the context of public service, though not specifically in Nigeria Public service in a particular implementation environment.

- Digital Transformation across the Globe: Introduction
- Examples of Digital Transformation Approach & Impact
- Digital Transformation Principles for Public Service

Chapter 4: Digital Government Trends

This chapter explores the digital transformation for a particular application within government and public services. It explores what digital transformation brings to this arena, and the environment and user experience it supports. The chapter ends by explaining what is needed for a successful digital transformation.

- Digital Public Services
- Defining Digital Transformation
- Defining Digital Transformation of Public Services
- Why Digital Transformation?
- Digital Transformation Essentials
- Chapter 5: Blockchain and the New Digital Economy

Chapter 5: Blockchain and the New Digital Economy

The chapter discusses the evolution of digital platforms into economic opportunity. The chapter explains blockchain and the bitcoin implementation. It focuses on the threat or opportunity for governments represented by blockchain. The chapter ends with an outlook for the new digital economy.

- Overview of Blockchain Economics
- The Development of Blockchain Technology
- Blockchain Transactions
- Strengths and Weaknesses of Blockchain Technology
- Blockchain Adoption & Future Outlook

Chapter 6: Digital and the Future of Government

The chapter discusses emerging digital technologies including an expansion on the blockchain. It focuses on the need for more attention to be paid to the understanding and the steps for implementing digital government services with an outlook on use cases:

- Digital Transformation and Blockchain Technologies

Section II: Strategy & Plan

Chapter 7: Public Service Digital Transformation

This chapter outlines the strategy and benefits of digital transformation to introduce a plan for Nigeria. The policy framework directly foreshadows the challenges of a Nigerian implementation. The chapter ends with an exploration of projected benefit reviewed in the context of the current Nigerian public service situation.

- Digital Transformation Strategy
- Key Benefits of Shared Technology Services in the Nigerian Public Service
- The Policy Framework for Digital Transformation of the Nigerian Public Service

- Digital Transformation with Current Public Service Reforms

Chapter 8: Digital Transformation Success Factors

The basis for an impact evaluation is presented in this chapter. Vision, investment, and leadership are highlighted. Each element is provided in the context of governance, infrastructure, and education supporting stakeholder buy-in at multiple levels of society.

- Envisioning the digital future for public service
- Investing in digital initiatives and skills development
- Leading the change from the top and achieving stakeholder buy-ins

Chapter 9: Digital Transformation and Nigeria

The challenges to the digital transformation that are unique to Nigeria are detailed in this chapter. The challenges are presented in a list. Each is explored and discussed before a response is prepared to utilise the benefit of digital transformation. The chapter ends by using the systematic advantage to build a case for digital transformation in Nigeria.

- Digital Transformation of the Nigerian Public Service challenges
- Defusing Challenges to Public Service Digital Transformation
- The Nigerian Public Service and the Case for Digital Transformation

Chapter 10: Digital Transformation and the Nigeria Approach

The basis for innovation is established in the Case for Digital Transformation. This chapter outlines the specific vision integrating digital transformation with the Nigerian approach. The method is discussed in three additional contexts with an application for other developing nations.

- Vision, Goals, Themes, and Guiding Principle
- Digital Transformation, Women's Rights, Poverty and Inclusion
- Digital Transformation and Sustainability
- Digital Transformation and Local Content Development

Chapter 11: Financing Nigeria‰s Digital Transformation

The financial, resource, and governance needs of the digital transformation are highlighted in this chapter. The chapter parallels the management, infrastructure, and education outline of the overall innovation while focusing on the financial aspects of the project. Management, infrastructure, and education are addressed in each of the sections noting that each is needed in Nigeria in the context of digital transformation.

- Essential components of Digital Transformation Expenditure: Outlines the complete Costs.
- Funding Sources: Identifies sources for funding the innovation.
- Leadership & Standards: Discusses the structure of a finance ministry to govern the change.

Section III: Implementation & Monitoring

Chapter 12: Implementing and Managing the Transformation Process

Presents the implementation process to be followed in the Nigerian project. It includes the focal areas organised into 3 phases: roadmap, agency, assessment, infrastructure, enablement, services, governance, and training. The process follows identified organisational change process: unfreezing, cognitive restructuring, and refreezing (Schein, 1996).

- Phase 1: Planning - Production of the roadmap, creation of the agency, and conducting of assessment.
- Phase 2: Innovation – Construction of infrastructure, enablement, and services.
- Phase 3: New Structures – Structuring governance and on-going training.

Chapter 13: Measuring the Transformation Progress and Impact

This chapter discusses the progress and impact indicators and their measurement. It focuses on two sets of indicators. The two include the process of innovation and the outcomes for the citizens.

- Requirements of the Impact and Output Indicators: Provides an overview of the indicators.
- Digitization Programme: Discusses the innovation project indicators.
- Population Outputs: Discusses the services, access, and opportunity indicators.

Chapter 14: Digital Transformation Key Performance Indicators

This chapter discusses the performance indicators and their measurement. It reports on three groups of output indicators. They include Infrastructure, Governance, and Institutional Capacity as groupings of indicators.

- Infrastructure: Reviewing the digital capability resulting from the innovation.
- Governance: Reviewing the digital governance structure resulting from the innovation.
- Institutional Capacity: Reviewing the overall services, efficacy, and education outcomes resulting from the innovation.

Chapter 15: Evaluation Parameters and Reporting on Nigeria Experience

This chapter discusses the reporting structure - combining the indicators into an annual presentation of gains, opportunities, and goals. The report communication strategy is detailed. The chapter ends with a projection of the training needed to maintain the progress.

- The structure of Reporting: Details the report outline.
- Communication Strategy: Maps the process for optimum dissemination of information on the innovation.
- Projections: Outlines the implementation schedule with the best chance for sustainability.

Chapter 16: Conclusion and Recommendation

This chapter recognises and summarised the enabling role of digital technology and provides governments with next step approaches for encouraging digitalization.

Epilogue

Author's Insight

About the Author

Copyright

Disclaimer

SECTION 1

Background & Service Review

"We don't see things as they are, we see things as we are"
— Anaïs Nin

Chapter 2

············◆◆◆············

NIGERIA: A BRIEF

"By 2020, Nigeria will have a large, strong, diversified, sustainable and competitive economy that effectively harnesses the talents and energies of its people and responsibly exploits its natural endowments to guarantee a high standard of living and quality of life to its citizens."

Mission statement: Nigeria's Vision 20:20:20

The Federal Republic of Nigeria is situated in Western Africa. It is the most populous nation in Africa and had a population of approximately 179 million people in 2014. [9] Its population growth rate is projected to be between 2.5 and 2.7 % per annum. The population of Nigeria is forecast to grow to about 310 million by 2035. The country has an average life expectancy of 54.5 years. [10]

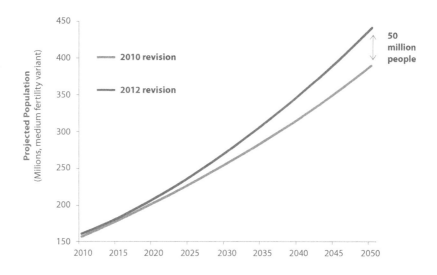

Figure 2: Nigerian Population Projection [11]

Despite still being defined as an emerging market, Nigeria is now the largest economy in Africa. It is the twelfth largest producer of crude oil in the world and has well developed financial and telecommunications sectors. Nigeria is one of 15 countries that comprise the Economic Community of West African States (ECOWAS) [12].

The enormous challenges faced by the people and government of Nigeria today cannot be overemphasised and may seem daunting. Some of these difficulties include infrastructure deficit, poor service delivery, security concerns, corruption, disregard for the rule of law and economic decline especially due to the recent downward trend in oil prices. 54.4 per cent of the population were living below the international poverty line of US$1.25 per day in 2011 according to the World Bank [11]. Nonetheless, a middle class has been seen emerging in recent years, especially in the larger cities of Lagos, Kano, Abuja and Port Harcourt.

NIGERIA'S POLITICAL SITUATION

After decades dominated by military leadership with the only inter-mittent civilian rule, Nigeria is being governed democratically since 1999. Its political system is a presidential democracy with a bicam-eral legislature composed of a senate and a house of representatives. Nigeria is a nation naturally blessed with human and natural resources in abundance. However, her political situation suffered a terrible stroke in the Nigerian civil war of 1967-70 which could be attributed to the intransigence of old politicians as well as the inability of the military to agree on the principle of justice, fair play and equity which was needed at that time. The African nation of Nigeria is an excellent example of post-colonial developing country, and its developmental history contains valuable lessons about the political and economic retardation of the developing world.

> *Nigeria offers, within a single case, characteristics that identify Africa. These opposing forces are rooted in the constant struggle in Nigeria between authoritarian and democratic governance, the push for development and the persistence of underdevelopment, the burden of public corruption and the pressure for accountability (Kesselman 515).*

Nigeria, as a country has witnessed six successful military coups, one violent civil war as well as the design of nine different constitutions. As the modern republic has adopted (supposedly), a model of federal democracy as a strategy to make sure there is national unity, still the consequence of many years of colonial and military rule persist, there is an emergence of a unitary system - *on with an all-powerful central government surrounded by mostly weak and also, economi-cally insolvent states.*

The country's party system has gone through a transformation process in which the People's Democratic Party (PDP) recently relinquished a great deal of its political power and is now the opposition after 16 years of uninterrupted rule. In February 2013, the All Progressives Congress (APC) now in power arose as the result of a merger of Nigeria's three biggest opposition parties at the time.

In April 2015, former military ruler General Muhammadu Buhari (APC) became the first opposition candidate to win a presidential election in Nigeria. He had previously ruled Nigeria from January 1984 until August 1985. His predecessor, Dr Goodluck Jonathan (PDP), led Nigeria between 2010 and 2015.

Recently, Nigeria has suffered several attacks by the Islamist militant group, Boko Haram, which is active in the North-Eastern part of the country. The group is responsible for the deaths of thousands of people and massive economic and social destabilisation in the region. It is also worthy of note that there have been recent attacks by militants in the Niger-Delta region against oil installations and pipelines. This, in fair measure, holds the promise of further destabilising the country's economy.

THE NIGERIAN ECONOMY

According to GIZ, the German agency for international development cooperation, Nigeria's Gross Domestic Product (GDP) was revised in 2013 and as a result of this re-basing, Nigeria's estimated GDP in 2013 went upward from 42.4 trillion Naira to 80.2 trillion Naira ($500 billion), an 89% increase. Thus Nigeria overtook South Africa as the largest economy on the African continent [13].

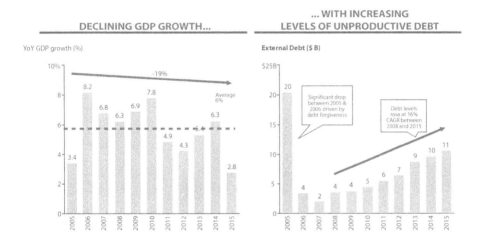

Figure 3: Nigeria's GDP and Debt Profile [14]

Since the 1970s, oil has been a dominant source of government revenues in Nigeria, but regulatory constraints and security risks have limited new investments in oil and natural gas with Nigeria's oil production contracting in 2012 and 2013. Even at that, the economy has continued to grow at a rapid 6-8% per annum, which is driven by growth in agriculture, telecommunications and support services. Economic diversification and strong growth have not translated into a significant decline in poverty levels - over 62% of Nigeria's 179 million people live in extreme poverty.

Nigeria's inflation rate has averaged 12% over the last 20 years and stood at 8.1% in 2014. The most significant sectors of the economy are agriculture, trade, oil & gas, information and communications, and manufacturing. These five areas represent more than 70% of the country's total GDP [14]. Its main businesses are oil & gas production, light industry (including aluminium processing, and paint production), food and beverage packaging, as well as subsectors that support the manufacturing of cement.

DIGITAL IN NIGERIA

Technology revolution in Africa has been steadily growing for the past couple of years as the rest of the world seem to be catching up with the continent's potential to be a major digital hub. However, when you talk about Africa, especially where technology is concerned, Nigeria will get recurrent mentions. Nigeria, as an emerging market, has undergone a genuine technological explosion of sorts. Start-ups keep popping up every other day, more and more investors are finding their way into the country, and technology adoption keeps increasing on a daily basis.

Nigeria is inevitably drawn into the global digital economy which is indexed by information and communication technology, given the rising knowledge intensity which permeates every sphere of life. Everyone has now come to the realisation of the fact that without the customer, no business would exist, and the customers are actively looking for businesses which would offer them innovative as well as value added products/services to better their lives. In this regard, placing target audience ahead of the competition is achievable via the digital evolution in Nigeria as it also positions industry leaders with unique offerings that the customers cannot do without. Digital is no longer the future; it is here Now! The combination of the elements of Internet marketing to deliver exceptional customer services and the social media marketing whose target audience is online are all the digital ways Nigerians have adopted over the years.

According to the "*we are Singapore*" digital portal, Nigeria has a population of about 179 million with urbanisation sitting nicely at 49%. Of the over 179 million people in the country, 97.2 million are active Internet users with an internet penetration figure set at 53% [15].

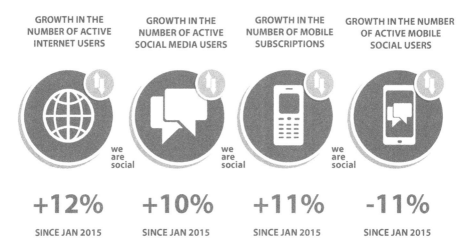

Figure 4: Digital in Nigeria [15]

Furthermore, their infographic shows that about 15 million Nigerians are active on social media while at least 84% of the population (or 158 million people) have an active mobile subscription (i.e. cell phones, SIMs, etc.). As for active social mobile users, Nigeria boasts a sizeable 10.0 million people with a 12% growth in the number of active internet users since January 2015. Nigeria also experienced a 10% increase in the number of active social media users, 11% increase in the number of mobile subscriptions, since January 2015. Either way you look at it, the future looks promising for the technology scene in Nigeria, and most of these numbers only prove that indeed, Nigeria has the potential to be the next big technology hub in Africa [15]

THE INTERNET IN NIGERIA

Internet Users in Nigeria (2016*) -	86,219,965
Share of Nigeria Population:	46.1 % (penetration)
Total Population:	186,987,563
Share of World Internet Users:	2.5 %
Internet Users in the World:	3,424,971,237

* estimate for July 1, 2016

In 2014, Internet penetration in Africa was only a 5% growth against the past year and increased from 13% to 18% in total. This was in comparison to a current global Internet penetration of about 46% making about 3.4 billion people. It should be noted that internet data from Africa are quite limited, and hence the growth in Internet penetration may be far greater than reported. Despite the tremendous increase in mobile penetration, in Africa and Nigeria in particular, digital content still has the lowest mobile penetration in the world at about 67%. Mobile broadband penetration in Africa is lagging behind the world at 17% but ahead of South Asia by 5%. Mobile broadband subscriptions in Africa have experienced tremendous growth and at 278 million, are ahead of Central America 81 million, South Asia 198 million, Central Asia 17 million, the Middle East 114 million and Oceania 35 million [15].

From 2016, social media penetration in Africa is expected to continue to lag behind the rest of the world at about 20% (tied with those of South Asia), but ahead of Central Asia at 6%. This is primarily as a result of the lack of affordable, accessible Internet on the continent. This represents a 3% increase from 2013. Nigeria minimally leads the way with the highest number of Internet users, Facebook users and active mobile subscriptions despite South Africa having higher Internet, Facebook and mobile penetration per population. The table below shows a rough breakdown of the numbers as elaborated by the International Telecommunication Union (ITU), World Bank, and United Nations Population Division [16].

Talking about the Internet in Nigeria, a worldwide computer network that can be accessed via a computer, a mobile phone, games machine, TV, etc. These services (internet) can be provided through a wired or mobile network, telephone line, ISDN (Integrated Services Digital Network), DSL (Digital Subscriber Line), or ADSL, cable modem, high speed leased lines, fiber, power line, satellite broadband network, WiMAX, Fixed CDMA, Mobile broadband network (3G, e.g. UMTS) via a handset or card, integrated SIM card in a computer or USB modem.

In February 2015, Nigerian Communications Commission (NCC) disclosed the number of Internet users on the country's telecom networks to be over 83 million. This data also revealed that there was an increase of 1,316,176 internet users on both the GSM and CDMA networks in February 2016.

Year	Internet Users**	Penetration (% of Pop)	Total Population	Non-Users (Internetless)	1Y User Change	1Y User Change	Population Change
2016*	86,219,965	46.1 %	186,987,563	100,767,598	5 %	4,124,967	2.63 %
2015*	82,094,998	45.1 %	182,201,962	100,106,964	8.4 %	6,348,247	2.66 %
2014	75,746,751	42.7 %	177,475,986	101,729,235	15.3 %	10,076,474	2.7 %
2013	65,670,276	38 %	172,816,517	107,146,241	19 %	10,487,424	2.72 %
2012	55,182,852	32.8 %	168,240,403	113,057,551	18.5 %	8,622,851	2.73 %
2011	46,560,001	28.4 %	163,770,669	117,201,668	21.7 %	8,298,063	2.73 %
2010	38,261,938	24 %	159,424,742	121,162,804	23.3 %	7,220,509	2.72 %
2009	31,041,429	20 %	155,207,145	124,165,716	29.5 %	7,074,482	2.71 %
2008	23,966,937	15.9 %	151,115,683	127,148,736	140.6 %	14,004,723	2.69 %
2007	9,962,224	6.8 %	147,152,502	137,190,278	25.4 %	2,015,189	2.68 %
2006	7,947,035	5.5 %	143,318,011	135,370,976	60.4 %	2,992,013	2.66 %
2005	4,955,023	3.5 %	139,611,303	134,656,280	183.2 %	3,205,447	2.63 %
2004	1,749,576	1.3 %	136,033,321	134,283,745	136.2 %	1,009,007	2.6 %
2003	740,569,	0.6 %	132,581,484	131,840,915	78.8 %	326,383	2.58 %
2002	414,185	0.3 %	129,246,283	128,832,098	265.6 %	300,896	2.56 %
2001	113,289	0.1 %	126,014,935	125,901,646	43.9 %	34,549	2.55 %
2000	78,740	0.1 %	122,876,723	122,797,983	60.1 %	29,565	2.55 %

Figure 5: Internet Penetration in Nigeria [16]

** **Internet User** = individual who can access the Internet at home, via any device type and connection

THE CIVIL AND PUBLIC SERVICE TODAY

The public service refers to all organisations that exist as part of government machinery for implementing policy decisions and delivering services that are of value to the citizens. It is a mandatory institution of the state under the 1999 Constitution of Nigeria. Public service is a service where federal, state or local government has an interest or where the presence of any of them is felt. In other words, it is a very large organ which encompasses civil service, the school system, the judiciary, the local government system, the security agents, and government companies. The public service is just of great importance to any given state. The bulk of decision and actions undertaken by the government are influenced and implemented by the public services. Furthermore, the public service will ever remain relevant in the scheme of politics and administration in Nigeria because it is well positioned to do so, but more importantly because of its long history, experience and specialisation [17].

In other words, public service includes the following:

a. The Civil Service: This includes the career personnel of the presidency state and local governments, ministers, extra-ministerial departments, the National Assembly and the Judiciary.
b. The Armed Forces, the police and other security agencies e.g. Para-military organisations.
c. The Parastatal or Public Enterprises

The Nigerian Public Service includes the Civil Service, often referred to as the Ministries, consisting of line Minister and extra-ministerial departments and agencies [18]. Most employees are career civil servants in the Nigerian ministries working in a progressive hierarchical system.

The civil service consists primarily of federal ministries, headed by a Minister appointed by the President of Nigeria while Directors and Commissioners lead civil service departments. A Minister leads each ministry in the Federal Cabinet to whom a Permanent Secretary reports. Some government functions are provided by commissions or parastatals (government-owned corporations) that may be associated with a ministry or may be independent, also sometimes headed by Permanent Secretaries, Chief Executive Officers, Director Generals, etc. The above scenario is replicated in the 36 states in addition to a total of 774 local government areas of Nigeria.

Public service is the vehicle or train that conveys service delivery and governance to the people, as the quality of any public service delivery can sometimes determine the rate at which the nation develops. The roles public servants play in the formulation and implementation of programs of governments cannot be overlooked, therefore.

The current Minister of Finance, Mrs Kemi Adeosun, speaking at a recent meeting with Newspaper Proprietors Association of Nigeria (NPAN) stated that the N165 billion monthly wage bill of federal civil servants was over-bloated and could no longer be sustained by the government, prompting calls for rationalisation or restructuring of the service overall [20].

THE HISTORY OF CIVIL SERVICE REFORM

Since 1945, various panels have been set up by successive governments to study and make recommendations for the reformation of the civil service. Popular amongst these were the Tudoe-Davies Commission of 1945, the Harragin Commission of 1946, the Gorsuch Commission of 1954, the Mbanefo Commission of 1959, the Margan Commission of 1963, the Adebo Commission of 1971, and the Udoji Commission of 1972-74.

Significant changes were undertaken based on the outcomes of the reports of the various panels which impacted on the structure of the service and the remuneration and productivity of civil servants. Also, the change from the Westminster model to the current presidential system of government over the course of time also impacted negatively on efficiency, effectiveness and productivity.

The Dotun Philips Panel of 1985 attempted to address observed lapses and inadequacies. The promulgation of the 1988 Civil Service (Reorganization Decree No.43) also had a tremendous impact on the structure and the efficient performance of the civil service. Nonetheless, at the inception of the current democratic system in 1999, it was observed by Faseluka Olugbenga in 'Civil Service Administration and Effective Service Delivery for Development' that, the public service and civil servants, in particular, were [21]:

- Lethargic and slow in making official decisions and implementing actions;
- Without a sense of urgency and were insensitive to the value of time;
- Frequently absent from work;
- Nepotistic, showing favouritism based on ethnicity, religion or politics;
- Wasteful of government resources;
- Poorly staffed;

- Corrupt;
- Inappropriately supervised;
- Characterised by a breakdown of the disciplinary system and code of ethics and
- Unresponsive and discourteous to the public.

Nigeria is one of the countries in the nations of the world which are currently in the struggle to better their existence through the process of good governance and responsible civil service for effective and efficient service delivery. There are certain factors that necessitated various reforms in the Nigerian Civil Service (1960-2005) owing to the situation of grand financial recklessness among civil servants which were facilitated by the long rule of the military and its attendant practices of impunity, lack of probity and accountability.

Subsequent study group reports such as the Alison Ayida Panel of 1994 not only reversed some of the structural innovations that had been implemented but also endeavoured to reinstate the historically held values of the civil service to no avail.

The Bureau of Public Service Reforms (BPSR) was established in February 2004 as an independent and self-accounting body. The mandate of the Bureau was to initiate, coordinate and ensure full implementation of government reform policies and programmes in the civil service [22]. Other significant efforts aimed at repositioning the public service to meet contemporary challenges as well as the aspirations of government were manifested in the previous administration's ''Public Service Reform'' programme. The programme's mission was to build "*a public service that is performance and results oriented, customer-driven, investor-friendly, professional, technologically sensitive, accountable, focused on fostering partnerships with all stakeholders and committed to a continuous improvement in government business and the enhancement of overall national productivity". Its overall vision was declared as: "A Nigerian Public Service that works efficiently and effectively for the people*" [23].

Despite substantive reforms across the country and its economy. Public debt is still rising (see figure 3 - Nigeria's GDP and Debt Profile), youth unemployment remains at disconcertingly high levels, public and individual health are still unfavorable, and Gross Domestic Product (GDP) growth is severely strained, reflecting the fact that the civil service is failing to adequately supply the much needed public services that would address those and other observed challenges. The present state of the civil service is that before and after independence, the Nigerian civil service has undergone several organisational and structural reforms, still there is no significant improvement in the quality of service provided to her citizenry as corruption still lingers and rears its ugly head again after so many years of military exit in governance. The national consciousness has been weakened due to earlier introduction of the "*Nigerianization*" and regionalization policies which encouraged ethnicity and sectionalism in the rank and file of staffers in the service. Today, of course, it is clear that these variables have not been totally dealt with, causing a failure to provide a panacea to the problem of development in Nigeria.

THE NEED FOR CHANGE IN
THE NIGERIAN PUBLIC SERVICE

The need for change is essential. An efficient operation of government bureaucracy is an important determinant of poverty, inequality, and standard of living. The public sector in Nigeria, is the nation's largest service provider, hence improvements in its operations and in its ability to serve the public, while also reducing corruption would positively impact millions of lives.

The nature of the services that the Nigerian public service is mandated to supply is at the heart of the need for public sector digital transformation. Healthcare, education, water and sanitation, roads and transportation, and electricity, are some examples of expected public services. Public services, in general, refer to those services that the government has a duty to provide, be it through direct state provision or otherwise, especially with regards to its obligations from a human rights perspective. In the context of a country with a large population living in poverty, few alternatives exist to, for instance, provide sanitation in rural or low-income areas, or in the delivery of maternal and child health care. The effect of the failure to provide these services is visible in Nigeria's human development indicators [24].

Public service delivery is the justification for the existence of the civil service and the ability of a government to legitimately tax and govern its people is premised on its capacity to deliver a range of services required by its population [23].

While the sheer size of the Nigerian population may be a governance challenge, the critical issues in question are a lack of prioritisation of citizens' needs, i.e., the service is not citizen-centric, and the misallocation of human and financial resources that follows, including an inequitable geographical distribution and a skewing of service provision towards urban areas [25].

It is undeniable that the Nigerian public service is in need of significant and encompassing transformation. Though it has been through such exercises in the past, as recent developments have demonstrated, the desired results have been slow to materialise. Moreover, the public's trust in the civil service has been eroded through years of endemic misappropriation, inefficiency and poor service delivery [24].

Government level discussions in recent years have justified the need for another round of reform in the service, to make it leaner, to improve capacity for efficient and effective service delivery and, crucially, to make its responses more dynamic to the needs of citizens.

Nonetheless, there are significant weaknesses in the service that have not been adequately addressed in these discussions, such as with maximising the skills, knowledge and motivation of its human resources. For example, there are some excellent leaders in the service, but not nearly enough of them, leading to the failure of many government programmes. Additionally, leadership in change processes needs to be made stronger and much more consistent, performance management is rarely rigorous, and the public service's culture has shown itself to be slow and resistant to change. In addition, movements across departments and ministries are rare: over 70% of public servants report never having moved organisations according to the BPSR in 10 years [26]. This lack of mobility in the labour market slows down the rate at which best practices for management spreads through the service and dampens incentives for organisations to tailor management practices to attract particular types of professionals.

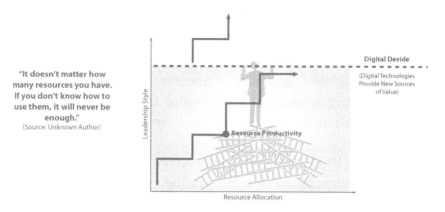

Figure 6: Government Resource Allocation

The current high cost of financing the Nigerian public sector is anti-thetical to sustainable economic growth. As the former Governor of the Central Bank, now Emir of Kano - HRH Muhammadu Sanusi II, rightly pointed out, the public service is currently over-staffed. Moreover, there is an alarming prevalence of ghost workers across over 251 Ministries, Departments and Agencies (MDAs) of government. The government also should as a matter of urgency extend its Integrated Personnel Payroll Information System (IPPIS) to all ministries, departments and agencies to attain a lean payroll and save on excesses and unnecessary expenditures [18].

Another compelling reason for reform is associated with identifying and eliminating wasteful spending, duplication and other inefficiencies across MDAs. For example, every public service organisation has a particular service or mandate to fulfil; however, it has been observed that for any given service, multiple organisations have been tasked to implement same in Nigeria. For instance, small-scale dams are constructed by the Federal Ministries of Water, Agriculture, and Environment, and by all of the River Basin Development Authorities – a total of 18 separate government entities [26]. Another example is the case of the Nigerian Communication Commission (NCC) and National Broadcasting Commission (NBC); Nigeria Police Force (NPF), Federal Road Safety Corps (FRSC), and Nigeria Security and Civil Defence Corps (NSCDC); Ministry of Works, and Federal Road

Maintenance Agency (FERMA) and many others as identified by the Oronsaye committee report [89] – a seven-member committee that was inaugurated by then President Goodluck Jonathan, on August 18th 2011. Major issues and trends that featured in the report about government Ministries and Parastatals under their supervision included the following:

i.   duplications and overlaps in mandates and functions;
ii.   requests for relocation and supervision of Parastatals;
iii.   top-heavy manning levels;
iv.   large and unwieldy governing boards;
v.   Parastatals/agencies without enabling laws;
vi.   Parastatals/agencies with expired mandates and tenures;
vii.   lack of inter-/intra-ministerial collaboration;
viii. inappropriate organisational structures;
ix.   the existence of agencies and professional bodies that should be self-sustaining but receive Government funding in addition to membership subscriptions;
x.   lack of clarity and transparency in the use of most generated revenue;
xi.   ineffective supervision of Parastatals by their relevant Ministries and divisional offices, which was exploited by "over-ambitious" Parastatals;
xii.   lack of appreciation and understanding as well as resistance to the intention of Government to streamline and restructure Government Parastatals and their governing boards;
xiii. lack of competence and quality in board membership; and
xiv.   the proliferation of research bodies across the Ministries.

Discussing further the core mandate of the National Identity Management Commission (NIMC), a successor of the department of national civic registration, the committee observed that there exist several government statutes that provide one form of identity data collection or the other. These they observed have created overlaps and excessive duplications in the public service. The relevant over-lapping statutes include the following:

1. The National Civic Registration Act;
2. The Passport Act;
3. The Immigration Act;
4. The National Population Act;
5. The National Health Insurance Scheme Act;
6. The INEC Act;
7. The FIRS Act;
8. Births and Deaths (Compulsory Registration) Act of the National Population Commission; and
9. The Pension Reform Act.

Whilst there was observed widespread malaise in the effective governance of Ministries generally, corruption and impunity in Nigeria's public service have become endemic. In 2012, Transparency International ranked Nigeria as the 35th most corrupt country out of 176 countries surveyed (bearing in mind that many other countries are tied on the same ranking). Much of this corruption is carried out in connivance with the public service. In 2007, Nigeria's corruption index revealed that the Nigeria Police, the Power Holding Company of Nigeria (PHCN), the Nigeria Customs Service and the Ministry of Education, (particularly higher institutions and examinations bodies) were the most corrupt public institutions in the country [18].

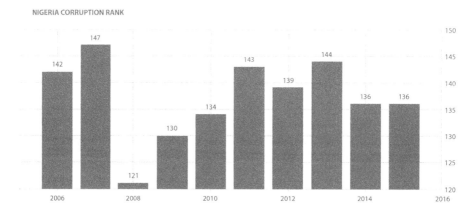

Figure 7: Nigeria Corruption Rank [27]

The latest pressing public policy issues prevailing in Nigeria, in times of harsh economic and budgetary pressures, is whether our public services will become more competitive and more efficient, i.e., to follow a lean government model. Rapidly developing information technologies such as Open Data, Open Procurement, Transparency and Enhanced Connectivity have the potential to revolutionise the public sector´s productivity, provide citizens access to services and usher a new era of participatory democracy, i.e., open and digital governance management. Lastly, issues of demographic change, employment mobility and the environment require public services´ immediate attention – a transformational government. Reforming public services may, therefore, contribute to economic and social development, fiscal consolidation, competitiveness and green growth prospects, as well as democratic governance [28].

The "*mentality*" of the public service is seemingly old and archaic. Thus, necessary graded training and exposure to new technologies should be done to facilitate the right mindset towards digitising Nigeria's public service. Indigenous technology companies with expertise in developing the technologies required should be used for the digitalisation process. National security considerations should be incorporated into the proposed transformation and digital government to minimise attack from hackers or rogue international actors.

DEFINING THE CHANGE REQUIRED

To address the challenges above and the need for change in the Nigerian public services, the following imperatives must be addressed:

1. The high cost of running the public service, including issues of ghost workers and overstaffing.
2. The proliferation of departments, agencies and Parastatals of government.
3. Bureaucratic, manual business processes and bottlenecks in service provision.
4. The lack of job security and the need for performance management.
5. Endemic corruption and a lack of accountability.

The service will require a new leadership model and much more sharing of services and expertise if it is to deliver the steep improvement and efficiency needed. It requires new service delivery models aimed at achieving better outcomes and lower costs for the government and it needs to transform its approach to the provision of public services to the populace, all of which can be facilitated by a digitally enabled transformation as proposed in this book.

As recently noted by Prof Yemi Osinbajo, Vice President of Nigeria, "*Certainly our country need a different set of values; a new way of doing business; an economy that can give opportunity to young people to work in their chosen professions and to build strong and profitable businesses*" hence this book [29].

"With Digital Transformation, the consumer,
rather than the technology, is in the driver's seat
and this matters"
— Forbes

Chapter 3

DIGITAL TRANSFORMATION LESSONS

The Digital Evolution Index (DEI) was developed by The Fletcher School at Tufts University in collaboration with MasterCard and Data Cash. The aim of the Index is to identify how selected countries around the world stack up against each other regarding readiness for the digital economy. The index is derived from four broad drivers: supply-side factors (including access, fulfilment, and transactions infrastructure); demand-side factors (including consumer behaviours and trends, financial and Internet and social media savviness); innovations (including the entrepreneurial, technological and funding ecosystems, presence and extent of disruptive forces and the presence of a start-up culture and mindset); and institutions (including government effectiveness and its role in business, laws and regulations and promoting the digital ecosystem). The resulting index includes a ranking of 50 countries, which were chosen because they are either home to most of the current 3 billion Internet users or they are where the next billion users are likely to come from [30].

The Digital Evolution Index

Index Score

0   50   100

Figure 8: Digital Evolution Index 2013

With DEI, Nigeria is ranked 50th in the world in 2013. Nigeria is classified as a 'watch out' country, which suggests that although we were rated at the bottom with smart approaches and knowledge, we can significantly overcome our limitations and challenges. Some combinations of emerging technologies - namely broadband connectivity, mobility, cloud, Big Data analytics, and social business – should sit at the heart of most public sector digital transformation efforts across Nigeria. While in the civil service, government MDAs are likely to keep increasing technology spending with little to show for the efforts as a result of systemic dislocation – as discussed in chapter two – Nigeria, A brief.

In Nigeria, many experts agree that digital commerce and retail are now the key and arguably the leading enabler for digital growth within the consumer industry in the country, adding that its electronic commerce is set to dominate African retail markets over the next five years. Retail sales have grown from $83.5 billion in 2010 to $124bn in 2015, and the Economist Intelligence Unit (EIU) forecasts the sector to reach $184.5 by 2019[81]. Consequently, a myriad of opportunities is being provided for consumers and businesses to buy and sell both new and used items online. Among such companies that provide such digital commerce, platforms are Dealday, Konga, OLX, Jumia, Kaymu, Wakanow, HotelNowNow, etc. According to KPMG fast-moving consumer goods sector report in Africa, Jumia and Konga are two big online consumer retail players, competing for market share with huge advertising campaigns. Retailers and telecoms providers are also teaming up; a 2014 deal between MTN and Shoprite allowed customers to purchase airtime and data bundles while shopping at Shoprite stores. Among diversified online retailers, there is an ambition to widen their retail activity to potentially include more FMCG activity. In 2015, Jumia incorporated over 150% more SME retailers and doubled the number of products available on its online marketplace [81].

DIGITAL TRANSFORMATION EXAMPLES, APPROACHES AND IMPACT

Africa's digital transformation is in full swing, and one of the places where lots of innovation is taking place is Kenya. In 2014, the Kenyan government unveiled its five-year Information Communications and Technology (ICT) Master Plan [31]. The purpose of the Plan was to review and update the Connected Kenya Master Plan earlier launched in February 2013 to (or "*intending to*") extending stakeholders participation while taking into account changes in the Jubilee Digital Government document. The plan recognised technology's critical role in driving the economic, social and political development of Kenya as espoused in its Vision 2030 agenda, and it focused on a roadmap to a knowledge economy and society that will lead to real socio-economic growth.

The Master Plan has three foundations and three pillars [31]. The foundations are the critical things that need to happen to lay a basis for Kenya's transition to a knowledge society and to position the country as a regional technology hub while the pillars are meant to facilitate the achievement of socio-economic growth and achieving its vision 2030 targets. The first foundation of the Master plan is technology, human capital and workforce development which aimed at developing quality technical personnel as a pre-requisite to the elaboration of a viable technology sector. The key to this is ensuring that technology development, implementation and exploitation are an integral and sustainable component of its national development. The second foundation is integrated information and communication technology infrastructure, which seeks to provide the integrated infrastructure backbone required to enable cost-effective delivery of technology products and services to Kenyans. The third foundation is integrated information infrastructure which aims at improving the quality of government digital services while enabling the country to transition to a knowledge-based society. This is done through ensuring maximum access to information knowledge held by public authorities by all Kenyans and by further ensuring that public information is readily available through a consolidated portal in an affordable and secure way.

The first pillar of this Master Plan is government digital services, which aims at ensuring the provision of digital government information and services as key to improving productivity, efficiency, effectiveness and governance in all key sectors. The second pillar is technology as a driver of industry, which aims at the transforming key for its vision 2030. This lies in developing the nation's economic sectors to enhance productivity, global competitiveness and growth significantly. And the third pillar is developing technology backed businesses that can produce and or provide exportable quality products and services that are comparable to the best in the world.

Once completed and implemented, it will completely transform public sector processes, services and management, and make information access and service delivery more efficient.

APPROACH

To implement Kenya's digital transformation, the appointed implementation task force began by reviewing existing technology plans, policies and guidelines in other to re-align with the strategic directions of the country. The final product became the National Information and Communication Technology (ICT) Master Plan for Kenya based on organisational and participatory approaches to strategic planning. The corporate approach ensured that key critical success factors and drivers had been fully aligned with digital transformation objectives as identified through context analysis while the participatory approach entailed taking care of the soft issues which are also critical for overall success. Examples of some of the soft key issues were attitudinal change, creating enthusiasm and motivating all stakeholders to rally behind the implementation.

IMPACT

Till date, the digitalisation efforts in Kenya are yielding many positive results and thus its citizens are better able to reap the dividends of digital transformation. The Kenya Civil Registration Department, for example, has succeeded as at January 2016 to digitise over 62 million birth and death records nationwide thus facilitating the easy search, retrieval of citizens' birth information at any location or county including the ability to query the database via SMS messages. Another major success recorded is that of the integrated financial management system which has completely revolutionised Kenya's public finance management from the point of planning through expenditure, execution and reporting thus enhancing transparency and accountability, according to Joyce Mugo, Director of Civil Registration in Kenyan government [32].

In Kenya, the road transport sector accounts for about 80% of the total road traffic in the country. The technology authority – the agency mandated to execute Kenya's ICT Master Plan and the National Transport and Safety Authority (NTSA) – are currently digitizing and modernizing the country transport industry by implementing the Transport Integrated Management System (TIMS), with the aim of promoting safety and centralizing all transported related data thus creating a single point of access to the entire transportation sector. The system currently supports over 500, 000 drivers with digitally enabled driver's licenses and SMS-based support and alerting systems including reminders for license renewals as stated by Fernando Wangila, Deputy Director, Head of ICT Directorate, NTSA [33].

It is envisaged that by the end of 2016 according to government officials, the integrated county revenue management system would have gone live in all of Kenya counties even though only about 57 locations nationwide currently enjoy the services of automated tax collection. The system is designed to ease the payment for services and collection of county taxes with the ultimate goal of feeding back into the national purse thereby guaranteeing complete transparency and visibility in the usage and collection of government revenue.

Finally, only about 47% of Kenyan citizens currently have direct access to electricity, a figure that Kenya Power vows to increase to 70% by 2017, and 100% by 2020. However, as Samuel Ndirangu, General Manager of ICT at Kenya Power and Lighting Company (KPLC) says, they cannot accomplish this simply by building more capacity. Most of the success will come from smart and lean operations that seek to gain efficiencies where they can be found. As an example, Kenya Power has adopted technology to cut down on costs by delivering electronic versions of billing statements to consumers. More than one million of Kenya's electricity users now receive their bills via email or SMS services. Secondly, KPLC currently spends most of its procure-ment budget with foreign source supplies and wants to support local industry by increasing procurement from local suppliers as a way to deepen local content development [34].

Safaricom, a subsidiary of Vodafone, the largest mobile network operator in the country, introduced the world's first mobile money soulition: M-Pesa. Safaricom users were able to easily adopt the service, because it was already installed on their SIM card [35]

According to Martha Chumo, founder Nairobi Dev. School, "*Innovative Digital Lessons from Kenya*" Kenya has emerged at the forefront of digital transformation in Africa, with many digital innovation examples and first [36]:

1. No smartphone? - Not just in Kenya, but in most of Sub-Saharan Africa (including Nigeria), most people use mobile phones to access the Internet. Not everyone has a smartphone, though. Individuals who want to access online services with a feature phone can use simple SMS application options. You can bet on your favourite team or get results on sporting events and information on job application via SMS. There are also SMS-based services for mothers who are pregnant or for mobile banking. Internet access is not a necessity for smart banking or online service interaction as already observed.

2. Mobile money is used for more than paying for stuff - In Kenya, for instance, you can pay for your taxi or drinks, electricity or TV bills with a mobile money service. But now, this has been transformed into banking. You can put money into a savings account and apply for a loan. This works well because not everyone has all the documents a bank needs when they apply for a microloan. But it's very easy to borrow small amounts through the mobile-based financial service M-Pesa, for example. You can also easily connect a mobile money service to your account at a formal bank and withdraw or deposit money there from your phone.

3. Digital technology facilitates social banking - Kenya has always had a culture of "*changas*", groups of people, men or women, who get together and instead of taking money to the bank, they "*bank

in each other". That is, they give small amounts of money to each other. There are now different services, such as Changasoft or M-Changa, which have moved this practice into the digital realm supported by mobile payment service.

4. Crowdsourcing is huge these days, but has a long tradition - Crowdsourcing is one of the easiest ways to get money today. But it's rooted in an old tradition that dates back before Kickstarter and Indiegogo were launched. If someone wanted to go to university or any school for that matter, villagers would get together and contribute money to allow that person to get an education. Now there's software that facilitates that. So for us, the idea behind Kickstarter or Indiegogo isn't a new concept, but the platform just allows us to do online what we've always done in person – she noted [35].

5. It's practical to use government services via social media - Authorities such as the immigration office or the higher education loan board are very active on social media in Kenya. They use social media to offer their services in more efficient ways. Officials who might not be able to serve everyone personally use Twitter to write short and concise messages. Their responses to inquiries from the public are also very quick.

6. Online services connect Kenyans with people in the diaspora - People living abroad like online stores such as MamaMikes to keep up the tradition of sending gifts to people still in Kenya, for instance, for Christmas. But more importantly, people in the diaspora can manage their investments back home or build a house via online services.

7. A plethora of apps offers useful, even life-saving tips - Apps have become a big deal in Kenya and provide a variety of services, including helping kids to do their homework after school. Another provides practical information to mothers on prenatal and postnatal care. These tools are very useful for women who live

far away from clinics and cannot easily go in for check-ups. The apps let mothers communicate with health care practitioners to let them know how they're doing and get medical advice in case they need help.

AUSTRIA

DIGITIZING THE COURTS SERVICE

Once the centre of power for the large Austro-Hungarian Empire, Austria was reduced to a small republic after its defeat in World War I. Following annexation by Nazi Germany in 1938 and subsequent occupation by the victorious Allies in 1945, Austria's status remained unclear for a decade [90]. A State Treaty signed in 1955 ended the occupation, recognised Austria's independence, and forbade unification with Germany.

A constitutional law that same year declared the country's "*perpetual neutrality*" as a condition for Soviet military withdrawal. The Soviet Union's collapse in 1991 and Austria's entry into the European Union in 1995 have altered the meaning of this neutrality. A prosperous, democratic country, Austria entered the EU Economic and Monetary Union in 1999.

Austria has been a leader in the use of information technology in Justice Management Systems in Europe. The transformation work started in 1986 with the introduction of a system for the mass processing of small money or debtor claims. In 1989 the Federal Ministry of Justice, in collaboration with the Federal Computing Centre – a centralised national technology agency, developed a system called Elektronischer Rechtsverkehr (ERV) – to allow the exchange of digital data between courts, parties and their legal representatives. The ministry continued its transformation efforts with the digitalisation of the land register, attestation register and commercial register, as well as the introduction of e-court-filing, electronic signatures and a video conferencing facility within the judiciary.

APPROACH

The digitalisation of Austria's judiciary and court systems followed a gradual approach and featured "carrot and stick" incentives to push the use of the new regime. Key elements of the strategy included the transition to a service-oriented architecture (SOA), the implementation of shared services and the standardisation of technology architectures across all judicial departments and agencies. A core outcome of the change was the automation of the court procedure system, which maintains a register of over 50 court processes, some of which (summary proceedings) are handled entirely automatically. The system allows new court entries to be transmitted electronically and court fees to be collected on a cash-free basis. Legal representatives can submit cases electronically, and courts can respond electronically. Reduced court fees were introduced to encourage the use of digital services. In 1999, a requirement was added for all law firms to have the technical facilities needed to support the system, and a year later communication through ERV became compulsory. In 2013, a new strategic initiative called Justice 3.0 was announced, with the objective of developing a vision for the justice system's whole technology landscape. It produced a roadmap for total digital transformation leading to the goal of an all-digital handling of judicial proceedings in Austria.

IMPACT

Austria is a global leader in the provision of information and communication technology (ICT) in the justice sector in Europe. Already the transmission of data is sent to the courts and prosecutors with the "electronic legal traffic" supported by automation. The electronic case management ("*Court Automation*") ensures rapid processing and storage of case data, including the creating and sending of documents, and also guarantees a process settlement in short periods.

The electronically guided land register and also the electronically governed Business Register, the Electronic Document Archive, the decisions database on the Internet and the introduction of video conferencing systems to reduce the number of legal aid interrogations complete the picture of a modern judiciary. Each employee of the Austrian justice system has a computer workstation and access to email and the Internet. In the network of the court are currently about 180 routers, 350 servers, 10,500 personal computers and 1,000 laptops connected. This measure not only increase the quality and speed of judicial power but also the job satisfaction of employees and has garnered numerous national and international awards for the Austrian legal system [38].

The Austrian government recently adopted a law that forces businesses to install electronic point of sale (POS) terminals which are linked to the Federal Computing Centre (Bundesrechenzentrum), the IT service provider of the Austrian Federal administration [35]

ESTONIA

PROVIDING DIGITAL GOVERNMENT AND SELF-SERVICE

Estonia, a minuscule Baltic country entered into the European Union in 2004. Estonia covers a territory of only 45,226 sq. km and there are only 1,342,409 inhabitants (data of 2007) – less than in some European cities and towns. Estonia's population ranks among the smallest in the world. It stretches 350 km from east to west and 240 km from north to south [37].

Estonia has been described by many as the world's most digital society. Its government was quick to embrace the digital economy, focusing on building an open society in the 1990s and introducing its "*tiger leap*" project to invest in technology infrastructure in 1996. The push for digitalisation continued through the millennium, and digital solutions are now at the heart of every citizen's interactions with the government. Estonians can vote electronically in general elections, file their tax returns online and sign legally binding contracts over the Internet. There are similar benefits for businesses: company registration is done online, and property deeds can be accessed digitally. Estonia's digital government has become a model for the rest of the world, giving citizens online access to information and public services and powering paper-free communication in the public sector.

APPROACH

Estonia's digital society was created not through a single overarching infrastructure, but through an open, decentralised system linking together multiple services and databases. The flexibility this provides has allowed new components to be developed and added over the years. All government digital services for citizens have a common user interface and a standard authentication system. Citizens and businesses conduct all their digital interactions with the government through one online website. Development began with the establishment of a functional architecture that contains the X-road system (a secure data transport system for government databases), the identity card and the public key infrastructure. Once these core services were in place, new elements were progressively added between 2000 and 2010: m-parking (mobile phone payments for parking), the e-tax board (electronic tax filing), digital signatures, an ID bus ticket, an e-government portal, i-voting, m-ID (a system for verifying online identity), e-police (a system providing patrol officers with a positioning system and a mobile workstation), e-health (digital health records) and e-prescription (digital prescriptions). The uptake of digital government has been aided by the popularity of the Internet (used by an estimated 81% of the population in 2014) and widespread support from public officials, businesses and citizens. In future, Estonia plans to integrate its service provision for all levels of government and offer a cross-platform self-service interface including digital data embassy [4].

IMPACT

Although there are variations between departments, the digital government has significantly increased the efficiency of public services overall. For example, registering a company now takes less than 20 minutes (reducing the time required to set up a business from five days to two hours), and more than 92% of tax declarations are made through the e-tax board, saving 7 per declaration. Also, the introduction of paper-free communications is generating significant savings across the public sector, with almost 2 million in savings for the Estonian Road Administration in 2011, for instance.

The 2007 cyber-attacks in which hackers compromised some government websites and services demonstrate that security still presents risks in Estonia. As more services are digitised and more people come to depend on electronic services (24% of votes in the 2011 parliamentary elections were submitted online compared to 2% in the 2005 municipal elections, for instance), security continues to be a priority [4].

FRANCE

DELIVERING A WHOLE-OF-GOVERNMENT TRANSFORMATION PROGRAMME

The country of France emerged from the fragmentation of the larger Carolingian empire when Hugh Capet became King of West Francia in 987. This kingdom consolidated power and expanded territorially, becoming known as "France". Early wars were fought over land with English monarchs, then against the Habsburgs, especially after the latter inherited Spain and appeared to surround France. French royal power reached its peak during the reign of Louis XIV (1642 – 1715), known as the Sun King, and French culture dominated Europe. Imperial power collapsed fairly quickly after Louis XIV and within a century France experienced the French Revolution, which began in 1789, overthrew Louis XVI and established a republic [90].

In 2007, France launched a comprehensive modernisation and transformation programme known as RGPP ("révision générale des politiques publiques" or general review of public policies). RGPP was intended to address three needs: improving the quality of public service delivery, cutting government spending by 10–15 billion by 2013 and continuing to modernise the civil service. The task of coordinating all RGPP initiatives was given to DGME ("Direction Générale de la Modernisation de l'État" or general directorate for the upgrading of the state), an inter-ministerial body established in 2005 by merging some existing agencies, which was to serve as the programme's delivery unit.

APPROACH

RGPP's overall mission was to serve French citizens better. After a spending review in each government department, opportunities were identified to save money and improve efficiency, leading to the selection of some 400 initiatives. They included structural reforms (including mergers between France's tax and collection agencies), changes in governance models (such as the implementation of a performance-based funding system for universities), service enhancements (for example, the acceleration of the naturalisation process), and improvements in support functions such as technology and human resource utilisation. To achieve the reform goals, the programme pulled a variety of growth levers, including lean operational techniques, information and communication technology, and performance management.

The DGME's role was to ensure that the digital transformation proved useful and that results were achieved quickly. Though from a range of backgrounds its entire staff had experience of conducting or supporting transformation projects. RGPP had support from the very top, with former President Sarkozy committed to the programme. By securing support from politicians and citizens at an early stage, it also obtained a mandate to deliver. It communicated its ambition and scope clearly to create a sense of purpose to which citizens could relate; equally, civil servants knew that the state had to modernise. There was pressure for an all-encompassing transformation across every administrative area touched by the RGPP so as to create momentum and make the process fairer. Progress towards reforms was communicated openly to the media and public to make the process of transformation highly transparent.

IMPACT

The reform programme has had enormous reach, involving all 2.5 million civil servants in France. By 2010, more than 7 billion of savings had been realised. Further reforms announced in that year were expected to yield an additional 10 billion by 2013. By 2011, RGPP had demonstrated tangible efficiency gains. Levels of service had been maintained even with 100,000 fewer full-time posts and surveys indicated that the complexity of public services as perceived by citizens had fallen by an average of 5 points since 2008.

The work of the DGME has led to a number of lessons for governments undertaking large-scale transformations: *secure support from the highest level of government; invite public scrutiny and be completely transparent so that citizens can see how the project is going; obtain tangible results quickly to reassure those involved that you are heading in the right direction; and invest in the skills needed to make the transformation happen* [4].

FRANCE

BUILDING INFORMATION TECHNOLOGY LEADERSHIP

As already mentioned above, the RGPP initiative was (see previous case study) to address three needs: improving the quality of public service delivery, cutting government spending by 10 - 15 billion by 2013, and continuing to modernise the civil service. The government decided to make Information Technology a priority for transformation because it served as a lever for process improvement and innovation, attracted state expenditures of 11 billion a year, and faced some challenges. These included a lack of transparency on costs, a decision-making process that was not aligned with business needs, poor vendor management, fragmented infrastructure, and limited talent management – all similar to the challenges in the Nigeria public service.

The government's transformation delivery unit, DGME, proposed to tackle these challenges through a large-scale interdepartmental IT project beginning in November 2009. In 2010, the government decided to invest 4.5 billion in developing networks, infrastructure and services to support a new digital economy.

APPROACH

In June 2010, France's Public Policy Modernisation Council appointed a Chief Information Officer (CIO) for the Government, charged with improving the coherence and interoperability of technology systems, promoting transparency and monitoring Information Technology costs and performance, controlling and mitigating risks in large-scale Information Technology projects, and managing or supervising cross-departmental projects such as infrastructure consolidation.

In February 2011, the CIO was put in charge of DISIC, a newly created directorate with some 20 staff responsible for centralising the French government's Information Technology services and improving their quality, effectiveness, efficiency and reliability. Its objectives included promoting IT cost controls among ministries, identifying savings in Information Technology procurement, rationalising national data centres by consolidating locations and infrastructures, and approving budgets for large Information Technology projects. DISIC also planned to centralise the rights governing the interoperability, security standards and accessibility of the government's Information Technology systems. By responding to the needs of government IT services in this way, it was hoped that DISIC would help to promote innovation and competitiveness in the wider public sector.

IMPACT

By 2017, DISIC is expected to save over 20% of the French government's annual spending on Information Technology, amounting to 1.1–2 billion. The savings are supposed to come from sharing infrastructure, purchasing and skills, promoting best practices in budgeting, and coordinating human resource management. Also, the CIO and DISIC have implemented some other initiatives including an IT maturity diagnostic, the establishment of a transparent IT budget for all ministries, a pre-launch project assessment, and a local Information Technology organisation [4].

SINGAPORE

ATTRACTING AND RETAINING THE BEST TALENT IN THE PUBLIC SECTOR

Modern Singapore was founded in 1819 by Sir Stamford Raffles (1781-1826). Raffles became a clerk for the British East India Company in 1795. He rose rapidly in the company. In 1805 he was sent to Penang and in 1811 he was made Lieutenant Governor of Java. In 1818 Raffles was made governor of Bencoolen on the island of Sumatra. Raffles believed the British should establish a base on the Straits of Melaka and in 1819 he landed on the island of Singapore. The island consisted of swamps and jungle with a small population, but Raffles realised it could be made into a useful port.

On 27 May 1961, the Malayan Prime Minister, Tunku Abdul-Rahman, proposed closer political and economic cooperation between the Federation of Malaya, Singapore, Sarawak, North Borneo and Brunei in the form of a confederation. The main terms of the confederation agreed on by him and Lee Kuan Yew was to have central government responsibility for defence, foreign affairs and internal security, but local autonomy in matters of education and labour. A referendum on the terms of the confederation held in Singapore on 1 September 1962 showed the people's overwhelming support for PAP's plan to go ahead with the merger.

Malaysia was formed on 16 September 1963 and consisted of the Federation of Malaya, Singapore, Sarawak and North Borneo (now Sabah). Brunei opted out. Indonesia and the Philippines opposed the merger. President Sukarno of Indonesia worked actively against it during the three years of Indonesian confrontation. The merger proved to be short-lived. Singapore was separated from the rest of Malaysia on 9 August 1965 and became a sovereign, democratic and independent nation.

Independent Singapore was admitted to the United Nations on 21 September 1965, and became a member of the Commonwealth of Nations on 15 October 1965. On 22 December 1965, it became a republic, with Yusof bin Ishak as the republic's first President. After that commenced Singapore's struggle to survive and prosper on its own. It also had to create a sense of national identity and consciousness among a disparate population of immigrants. Singapore's strategy for survival and development was essential to take advantage of its strategic location and the favourable world economy.

In the 1990s, Singapore recognised that it needed to reform human resources in the public sector for two reasons. First, globalisation and global competition for talent were changing employee demographics and attitudes. Second, there was still a legacy of some practices from colonial times, such as a system of promotion based on seniority as is the case in the civil service of Nigeria. As the largest employer in the country, the government also faced the challenge of coordinating, integrating and managing an increasing number of agencies. In May 1995, the government launched 'Public service for the 21st century', a transformation movement to encourage officials to embrace change in their daily work and a platform for supporting an organisational change in the government itself.

APPROACH

In 1995, the government devolved Human Resource Management from the Public Service Commission into a system of personnel boards. The Public Service Division, Singapore's central HR agency for public services, has developed a meritocratic framework to appoint and develop civil servants who are collaborative, service-oriented and strong performers. The government gives high-flying school students full scholarships to attend top universities, in return for which they must work for the government for several years after graduation. Existing members of the civil service can apply for sponsorship for postgraduate study as part of their career development. Singapore's public service also has a strong focus on training, with officers receiving 100 hours per year. A dedicated training college, Singapore Civil Service College, offers more than 150 courses. To ensure that the public sector retains a fair share of the nation's talent, civil service pay is also at the market rate, with salaries comparable to those of private-sector employees with similar abilities and responsibilities. Pay is also linked to performance, with bonuses to reward high achievement. The system is being adjusted so that pay progression is based not on fixed annual increments but potential and performance assessments. Periodic salary reviews are held to maintain market competitiveness with the private sector. For example, there was a 5% pay increase across the board in August 2014.

IMPACT

Regarding its success at retaining and cultivating talent, Singapore's civil service has been hailed as a model for the rest of the world regarding meeting citizen's needs, despite spending levels as low as 19% of GDP. In a recent study, 56% of respondents expressed satisfaction with Singapore's public services, compared to a global average of 36%. Measured by the World Bank's Worldwide Governance Indicators, Singapore ranks almost best-in-class among countries for effectiveness, regulatory quality, the rule of law and corruption control [4].

Global Financial centers index, singapore is ranked the 4th largest financial center in the world, and is also ranked in 1st place on the World Economic Forum Global Information Technology Report 2015. This confluenece of financial maturity amnd technological readiness is critical for the accleration of the Fianacial Technology industry [35]

SWEDEN

MANAGING THE BUDGET CRISIS

After the death of the warrior king, Karl XII in 1718 and Sweden's defeat in the Great Northern War, the Swedish parliament (Riksdag) and the council was strong enough to introduce a new constitution that abolished royal absolutism and put power in the hands of parliament. Eighteenth-century Sweden was characterised by rapid cultural development, partly through close contact with France [90]. Overseas trade was hard hit by the Napoleonic Wars, which led to general stagnation and economic crisis in Sweden during the early 19th century. At the end of the nineteenth century, 90% of the people still earned their livelihoods from agriculture.

One consequence was emigration, mainly to North America. From the mid-19th century to 1930, about 1.5 million Swedes emigrated, out of a population of 3.5 million in 1850 and slightly more than 6 million in 1930. The industry did not begin to grow until the 1890s, although it then developed rapidly between 1900 and 1930 and transformed Sweden into one of Europe's leading industrial nations after World War II. Sweden is one of the largest countries in Europe, with a great diversity of its nature and climate. Its distinctive yellow and blue flag is one of the national emblems that reflect centuries of history between Sweden and its Nordic neighbours.

In the early 1990s, the bursting of a real estate and financial bubble tipped Sweden into a deep recession. Between 1991 and 1994, the national debt doubled, and unemployment soared from 2% to nearly 10%. The recession combined with a weak market for state bonds to push up interest rates sharply, driving up the cost of debt maintenance. As a result, Sweden's budget deficit grew rapidly, reaching 10% of GDP, the largest of any OECD nation, by 1993. When a new Social Democratic government came to office in October 1994, it launched a programme intended to stabilise national debt as a share of GDP by 1996, reduce the public finance deficit to a maximum 3% of GDP in 1997, balance public finances in 1998, and achieve a public finance surplus after that.

APPROACH

The devaluation of the Krone by 20% in 1992 had already helped in the management of the budget crisis by boosting growth and kick-starting the economy. The budgetary reforms were designed to strengthen public finances by increasing government receipts and reducing expenditure. Between 1995 and 1998, the government implemented a series of tax rises and spending cuts affecting almost every area of the economy. In total, half of total savings came from tax and fee increases, and a half from expenditure reductions. To increase its income, the government raised taxes on capital and capital gains, general employment, state income, and share dividends, as well as raising employers' health insurance fees and reducing deductibility on pensions. To cut expenditure, the government reduced housing subsidies, child allowance, child care subsidies and compensation levels in social and labour market insurance. Also, most government agencies were required to make one-off savings of around 10% of expenditure. The government then introduced a forcing mechanism for productivity in which each government agency received by default the same budget in nominal terms every year, requiring it to make savings of up to 3% per year to counter salary inflation. This policy helped to improve efficiency and innovation in service delivery. The reform had five cornerstones: to obtain a reality check through a thorough analysis supported by transparent and public discussion; to construct a programme that both offered a solution to the crisis and included safeguards to prevent any recurrence; to obtain a mandate by involving parliament, trades unions, employers' federations, the media and specific individuals and organisations; to ensure that the administration was motivated and had clear targets and freedom to act but was accountable for its actions; and to shift to a new agenda by planning more positive reforms once the austerity measures had achieved results. Another important goal was to maintain a sense of social justice in who bore the effects of cuts and service reforms. For example, the highest tax bracket was temporarily raised from 53 to 58%.

IMPACT

The programme achieved the government's goals: national debt as a percentage of GDP was stabilised, the deficit was reduced to 3% by 1997; and public finances were balanced, with a surplus of 1.2% of GDP in 1998 and 1999. Disposable household incomes initially fell between 1994 and 1997 but picked up again in 1998 with 2% growth as a result of public sector transformation [4].

EUROPEAN BANKS

INTRODUCING STRAIGHT-THROUGH PROCESSING TO REDUCE COSTS

The analysis shows that applying digital tools and methods to change the way banks process and service customers can reduce their cost base by up to 25%. Much of this potential saving comes from the automation of manual tasks and the introduction of straight-through processing (STP), which eliminates lengthy and complex manual processes. Through the deployment of workflow tools and self-service capabilities for customers and staff, STP has the potential to reduce the costs of a range of internal processes by up to 90%.

APPROACH

The financial services industry is a leading practitioner of the STP approach and is undergoing digitalisation at a rapid rate, although most banks face considerable obstacles, including a highly complex business context, a legacy of IT architecture and out-dated IT development capabilities. To achieve STP in a business process, banks need to follow four steps: prevent paper input by the customer; digitise the flow of work to enable automation; support the decision-making process through the use of software and analytics tools, and improve the productivity of remaining work such as front-end operations. Banks typically begin by reinventing a given process from scratch, rapidly creating a *"minimum viable product"* and developing the process iteratively with continuous customer testing. By combining this rapid end-to-end digitalisation with lean and agile development methods, banks can digitise a process in 16 weeks.

Banks that have been successful in automating their operations share three factors in common:

- Working with the business to simplify the existing process. This requires a cross-functional team of operations, technology and business experts with strong project governance and top management support.
- Using multiple integration technologies and approaches. Banks can automate most manual interventions without having to rewrite or change their existing IT architecture, but by using the right mix of integration solutions instead.
- Prepare the IT shop for agile development methods. To build business enablement skills, banks should hire people with expertise in applying the right solutions or provide appropriate training.

IMPACT

Straight-through processing has delivered cost and time savings in many different banking functions. One European bank automated its account-switching process and achieved a return on investment of 75% after just 15 months. Back-office staff handled account change-overs far more quickly, saving 70% of the processing time, while the time it took customers to switch was reduced by more than 25% [4].

Another European bank wanted to shrink its cost base and boost its competitiveness by offering a superior customer experience. Its auto-mation programme focused on ten major processes including retail account opening and wholesale customer service. The programme became self-funding in the second year of implementation, and the bank gained some business and operational benefits.

OFID

TRANSFORMING OPERATIONS

The OPEC Fund for International Development (OFID) is the development finance institution established by the Member States of OPEC in 1976 as a common channel of aid to the developing countries. OFID works in cooperation with developing country partners and the international donor community to stimulate economic growth and alleviate poverty in all disadvantaged regions of the world. It does this by providing finance to build essential infrastructure, strengthen social service delivery and promote productivity, competitiveness and trade. OFID's work is people-centred, focusing on projects that meet basic needs - such as food, energy, clean water and sanitation, healthcare and education – with the aim of encouraging self-reliance and inspiring hope for the future.

In OFID, the Public, Private, Trade and Grant operations units are charged with the onerous responsibility of delivering on the institution's noble mandate and her strategic mandate of "Uniting against Poverty". As the Fund's operational base continues to grow and expand due to increased portfolio potential, and as new windows of business opportunities are being exploited, the imperatives for a more integrated, standardised, reliable and secure enterprise information management system becomes even more critical. Digital advances such as analytics, mobility, social media, cloud, Internet of Everything and big data will continue to underpin and support the evolution of OFID's global operations and further enhance its avowed commitment to help end extreme poverty.

APPROACH

Implementing a solution at the enterprise level allows the Fund to get rid of standalone and disparate computer application systems, replacing them with a single enterprise software solution with individual specialised sub-modules dealing with varied needs. The key difference is the creation of a single version of *"business truth"* thus eliminating enterprise data inconsistencies and organisational conflicts. This requirement is now a common denominator and country independent legal statute for both national and international firms and transactions. The approach was to:

- Eliminate manual procedures and optimise core operational processes, thereby fulfilling operational and accounting compliance for all operations.
- Centralise and standardise business partner management to achieve alignment and synergies across the institution.
- Integrate an Electronic Document Management System with backwards enterprise compatibility and integration with its ERP system.
- Centralise and standardise record archiving and storage with necessary security and accessibility controls.
- Enable more efficient decision making and reduce costs over the long term through effective reporting and proactive business intelligence [40].
- Support effective and efficient corporate planning through consistent data provisioning and standardised reporting.

IMPACT

At completion, the digitalization efforts reduced the institution's human and technical complexities, minimising work effort and operating cost. The new system continues to demonstrate its benefits and the operational efficiencies of running a shared services centre of excellence, particularly for OFIDs' core operation and other support services. It also completed the provisioning of the requisite integrated technological infrastructure and service base required to support OFID stakeholders' community and also deliver significant improvement to the attainment of its noble mandate. [40].

DIGITAL TRANSFORMATION PRINCIPLES
FOR PUBLIC SERVICE

Digital transformation of the public service, stemming from the above global lessons should operate on the following guiding principles for effectiveness:

1. Open data and open interfaces leading to an open government.
2. The efficiency of technology services and infrastructure, based on sharing of Platforms, Applications, Components and Services.
3. Efficient business processes, stressing the need to optimise existing business processes before digitising them.
4. The quality of digital services, emphasising user-centricity and experience as a requirement, as well as the security of digital services by design.
5. Equal access to digital services; ensuring that public services provided are equitable despite spatial, economic and social differences.
6. Tailor-made solutions for rural and urban areas that recognise the different characteristics of these environments.
7. Stakeholder awareness and participation, and a pro-poor approach.
8. A multi-sectorial perspective and the need for cost optimisation.

Case Study: Kiambu County emerges the leading county in the Kenyan Government's Digital Transformation

Kiambu County is leading in the Kenyan Government's digital trans-formation drive with the recent launch of a new digital platform. The Kiambu Digital Platform includes the enablement of cashless payments, an online citizen services portal, and a robust back-end management system to support it.

Through the system, the country will address five key challenges that include collection of revenue to bridge the revenue collection gap, improve service delivery to the citizens of Kiambu, higher fiscal accountability for the people of Kiambu, better control over expenses and sealing 'leaks,' (i.e. reducing waste). The Kiambu County government is also looking to create 50,000 jobs through sustainable economic initiatives within the next five years. The innovation follows the identification of the challenges back in 2013 which led to the idea of embracing technology. "The first step of meeting these challenges was to implement a new system that would allow for better revenue collection. Our existing collection systems were consuming most of the revenue, leaving us with little to apply to much-needed development programs. Since implementing the new revenue collection solution, we have freed up budget and resources that were tied to internal operations, leading to the county recording a 90% growth in revenue collection for the financial year 2014-2015." Kiambu County governor explained [37].

"The digital revolution is far more significant that the
invention of writing or even printing."
— Douglas Engelbart

Chapter 4

...........•◆•...........

DIGITAL GOVERNMENT TRENDS

The Accenture Technology Vision 2015 highlights five trends that are reshaping industries and changing the way people live and work around the globe. The trends foresee a significant shift and opportunities to improve public service performance and citizen satisfaction with government services. They point to the government's unique role in shaping and enabling the new digital economy with infrastructure investment, digital-friendly policy and public/private partnerships that stimulate innovation. Ultimately, they reinforce the idea that very soon every government will need to become a digital state [42].

Trend 1: Internet of Me - The Internet of Me is changing the way people around the world interact through technology, placing the individual at the centre of every digital experience. Whether checking bank accounts, listening to music or monitoring our health, we now expect intuitively, at-our-fingertips digital experiences that reflect our preferences and address our needs. Such services offer distinct benefits to consumers, but they can also provide a powerful channel through which service providers can influence consumer behaviour. The push for consumer-centric digital experiences is both raising the bar and fuelling opportunities for digital government. When well executed, a seamless digital experience can lead to more than just satisfied citizens. It also can help government encourage positive behaviours such as greater voluntary tax compliance.

As digital becomes even more embedded, they expect radical shifts in the way people live and work. Consider, for example, the on-going shift from supporting "*health and welfare*" to emphasising "*well-being.*" Digital technologies can be used to provide personalised support toward healthier lifestyles, enhancing individuals' personal growth and their contributions to the community.

That kind of personalised digital experience requires much more than a refreshed portal. Digital governments must know the needs, moti-vations, preferences and pain points of their "customers". They also must own and shape the entire "customer" experience. Ultimately, providing a personalised experience requires digital governments to achieve a higher level of sophistication at every level and in every aspect of their operations—and to apply that standard consistently across thousands of citizen touchpoints.

Trend 2: Outcome Economy - A growing collection of connected devices are changing the way we live and work. As objects come online, they enable new services. For example, sporting goods companies now promote fitness services enabled by personal well-ness devices that capture and share exercise data. This is only the beginning of an explosion of new devices and new ideas fuelling an Outcome Economy. Sensors, cameras, household appliances and vehicles are among the myriad of things now connected to the Internet. For each connection, there is the potential for new services and better decisions on everything from how to get to work to where to go for dinner. This Internet of Everything will enable organisations to offer enhanced services, such as remote vehicle maintenance, property management and personal health management.

Digital devices will tie together businesses, governments and individ-uals from every industry of the world. In time, this will transform the way governments engage with citizens and deliver outcomes. Smart parking systems—which help drivers more quickly locate available spots—provide an early example of connected devices supporting an outcome-focused approach in the public realm. Likewise, connected and intelligent cameras and sensors can help improve public safety outcomes by providing officers with real-time situational data, such as gunshot detection and alerts of suspicious behaviour. The ultimate opportunity is for individuals to personalise their environments and, thus, their ability to address their own unique needs. The more specific opportunities of digital devices are only just beginning to be understood, and we have a long way to go as the digital revolution continues.

Trend 3: Platform evolution - The Platform evolution reflects how digital platforms that facilitate exchange, aggregation and analysis of data are becoming the tools of choice for building the next generation of products and services. A precipitous drop in the cost of data storage and computing power is enabling a level of data sharing between an organisation that was previously unimaginable, and it is giving rise to new digital "*ecosystems*" which connect stakeholders, service providers, partners and vendors in new and unconventional ways.

Digital platforms are at the heart of the on-going digital revolution. They provide the place for collecting, sharing and aggregating data, performing analytics and delivering new and improved services. Digital platforms help eliminate traditional data-sharing barriers between organisations—connecting all of the providers and people who collaborate to provide a service. On the commercial side, platforms may connect all members of the product supply chain—from raw material suppliers to end customers. But this is just the beginning. Wherever there is value to be gained by connecting organisations (or "*governments*"), digital platforms can open new possibilities. For example, power consumption data from smart building management systems can be shared with local power utilities to forecast demand better and reduce energy costs for a connected city. Today, organisations often leverage cloud-based, as-a-service platforms to substantially reduce start-up costs while allowing for rapid expansion as demand grows.

They envision that the digital government of tomorrow will include digitally enabled ecosystems in each of the mission areas of government (for example, education, health, social security, public safety, economic vitality, quality of life and smart cities). These ecosystems will connect not only the government agencies but parties such as NGOs and the public, too. Successful digital ecosystems will be those with strong governance, standards and platform–enabled interoperability. As an example, consider that in any country, the health ecosystem is one of the largest and most complex. Digital platforms supporting a connected health ecosystem can drive significant benefits much faster and at a lower cost—from more easily identifying and booking a doctor's appointment to improving the transition of a patient from a hospital to a community care setting. Implementing such services for the entire ecosystem promises to optimise utilisation of costly care facilities while improving the personal health outcomes for the patients being serviced by the system.

While the Platform evolution may be a longer-term reality, platform evolutions are underway. Health, public safety and human services organisations are deploying digital platforms that span multiple jurisdictions to help coordinate service delivery. For example, many public safety and criminal justice ecosystems share information about offenders or other persons of interest. As these ecosystems evolve, they are moving into real time and are enabling analysis and dissemination of data to and from a greater range of sources. Likewise, open government platforms represent early instances of information marketplaces that combine public and private sector data. Developed by a variety of entrepreneurs, these marketplaces can support a new generation of services. At the same time, digital governments continue exploring the "as a service" delivery options for their core systems. This shift away from on-premise solutions is crucial to help lay the foundation and build momentum for adopting next-generation digital technology platforms.

Trend 4: Intelligent Enterprise - The trend toward an Intelligent Enterprise is about embedding software intelligence into every aspect of a business to drive new levels of operational efficiency, evolution and innovation. In the Intelligent Enterprise, insights aren't trapped in offline systems or delivered to only a handful of workers. Instead, the Intelligent Enterprise spreads insights throughout the organisation, enabling a greater number of employees to make more—and more efficient—decisions. Intelligence may be embedded in infrastructure (such as utility networks, operational machinery, buildings and vehicles) or within business processes (using software intelligence to help automate steps or provide recommendations and better decision-making insight to the worker). Embedded software information can help governments identify underpayment of taxes and, using diverse data sets, find clues to determine the next best action.

Other applications of embedded intelligence include enabling operational efficiency, tracking policy effectiveness in achieving target public sector outcomes, and suggesting changes to be made. Indeed, for digital governments, the opportunities and implications are enormous. With better software intelligence and data-driven insight, digital leaders can increase effectiveness in delivering a range of public service outcomes—from better public safety and social services to greater economic vitality. Getting there requires a cultural shift in the organisation, whereby machine intelligence is seen and embraced as an aid in decision making rather than as a threat to workers themselves.

Trend 5: Workforce Reimagined - Advances in more natural human interfaces, wearable devices and smart machines are extending the intelligent technology to interact as a "*team member*," working alongside employees in a Workforce Reimagined.

The nature of work is changing. Already we have moved from lifelong employment to periodic employment. The future is poised to bring crowd-sourced employment where work could come from anyone and anywhere. Digital assistants, wearable devices, robotics and digital printing all will supplement people in the workplace, enabling higher productivity and changing the way that works gets done. The future will also bring greater focus on digitally enabled access to and sharing of skills and experience. Governments will scale and extend the use of digital mentors, remote working and job sharing—all with the aim of improving productivity, personal growth and well-being. Importantly, Workforce Reimagined isn't about drones replacing humans. Rather, it's about stretching the boundaries of technology far beyond mere automation—applying a new generation of technologies to augment the cognitive, collaborative and physical capabilities of human beings. In short, digital leaders will use technology to make their workforces more productive, more efficient and happier. For digital governments, a reimagined workforce is one with strong, peer-to-peer collaboration and insights available in the field at the points of service or need. To make the best use of these advances, workers must have the authority to act on ideas with minimal "red tape." For most governments, getting there will require significant changes in both governance and culture as part of the journey to becoming a digital society.

All peoples' expectations are being steadily shaped and influenced by their experiences across the numerous other technology platforms, from online shopping to technology. A truly digital government can help meet and satisfy these growing expectations. A global scan of leading citizen-centric practices reveals that governments globally can do more by harnessing the right technology to build trust and transform.

Similar to Accenture, PwC has identified seven global trends for digital government in their latest report as follows [43]:

Think Digital First - a digital-first mindset lies at the heart of faster and better service delivery. As already discussed in the book, the Nigerian government may soon find itself struggling to meet up with the pace of digital technology and the expectations of its mainly youthful population. Some of the things citizens want are consolidated identification systems – whether its passports, driver's licenses, voter registration cards, etc. and information as well as digital applications and approvals.

Mobile Now – The rapid adoption of technology is now disrupting the way people consume digital services. It is possible to access services on the go through a mobile device, from submitting applications to accessing information and government services should be no exception. A mobile government provides services that add value to the lives of their citizens including reminders like permit expiration, voting dates and locations, etc. digitally.

GSM Association expects an additional half a billion mobile subscriber and 240 million mobile internet users in sub-saharan africa over the next five years[44].

Omni-Channel - By enhancing multichannel capabilities, governments can provide services through various channels including mobile, website, offices, phone and even watches. Applications can be started and fulfilled on different channels. An integrated information centre is one element of a multi-channel approach to digital government. It can build on existing technologies and information from several government MDAs acting collectively as a single government information portal.

Leveraging Citizen Data – In a properly implemented digital government, supporting collaborative and innovative solution platforms, governments can take advantage of citizens' data to provide the right services proactively. An example is seen from Singapore's My eCitizen portal which provides citizens with the option of receiving proactive notifications through text messages from the government on diverse matters from library book reminders to passport and license renewal.

Tracking and transparency - People are demanding to stay informed and want to know the status of their applications and permits on the go. Allowing access to this information via system integration will increase transparency and improve service delivery. It will also reduce calls and visits to government offices. In Denmark, the government has provided a personalised central portal where citizens can access all their pending transactions with any agency.

Cyber security - Governments by their nature of operation, can become holders of critical information, and by this information, citizens security, especially cyber-security can be enhanced or attacked. Today, risk, terrorism and cyber-attack incidents are on the rise. Protecting data and reassuring citizens that their data is secure are critical. Estonia improved its cybersecurity policies and practices in October 2008 after the denial-of-service attacks in May 2007. Improvements included setting strategic goals and measures, recommending regulatory and legal framework adjustments and emphasising international coordination to address cyber threats.

Efficiency is key in government - To give citizens the digital and quality service delivery they're looking for, operational excellence and efficiency are essential, supported by the right technological tools. United Kingdom's London Borough of Harrow has been working through its Access Harrow program to improve access to council services and add greater levels of process automation across channels, including telephone, walk-in, paper and web. As an example, a case study on an innovative title search application is presented in this book.

DIGITAL PUBLIC SERVICES

A digital service is a term applied to public services that are delivered electronically over the Internet. Such services are typically cost effective as they have a little substantive dependence on staff or a physical presence. They exist only within a digital environment.

Eddie Copeland says, "*Technology opens up the prospect of government being in a position where it doesn't have to do everything, but can act as a coordinator to make things happen. We now use technology to collaborate in so many other aspects of our lives; I'd like to see more of this in the public sector*". In developing countries now, there are a lot of great digital initiatives with so many of the best innovations in the space springing up in countries in Africa. Also, most successful examples close the feedback loop by working with the government in some way, and the private sector is quicker on the uptake when the government are much slower than them in the drive towards digitalisation. Basically, from experience, it is needed to start with the needs of the citizen and build the service from that point backwards into the government organisation providing the service. Meanwhile, because efficient digital services are not about technology, rather they are about empowering people and giving them a say in what was once the sole purview of public service administration.

"*The biggest, most earth-shattering fact to understand, and one that is worth repeating, is that the unbanked are being reached. The cash economy is being supplemented by mobile access to digital funds. Traditional banking is not leading the change. Telecommunication companies have stepped in and mobile powered transactions are the force to be reckoned with. Disruptive? Yes. Transformative? You bet*" [35]

The digital environment includes a range of digital channels and a large number of digital touch points. Digital channels include the web, mobile applications, smart interactive voice response (SIVR) applications, emails, social media, web chats and SMS (text messages). Digital touchpoints usually include specific websites (in the web channel) and mobile applications (in the mobile application channel). Touch points may contain digital content such as information about a service and transactional capability, for instance, an online form. Also, the digital environment can also include back-end digital service delivery infrastructure that customers can't see and don't interact with directly.

Thus, for the purpose of this book, we can derive the definition of a digital public service as *'a public service delivered digitally over the Internet that offers a simple and accessible way for citizens to transact with the public sector [45].*

Digital public services aim to be simple and accessible, and available to be consumed at times and locations that suit citizen preferences. These services are transactional, meaning that digital public services are typically more than just informational and the critical aspect is a new or improved transactional capability.

DEFINING DIGITAL TRANSFORMATION

Public sector administrators must tackle many new challenges, including those regarding the demographic change, employment, mobility, security, environment, food security, among others. The recent technological initiatives such as open data, open source and take up of social media has led to more information and knowledge exchange as well as enhanced connectivity, openness and transparency on all levels [46].

The essence of digital transformation can be understood as an organisational change process in which digital technologies (such as social media, telecommunication, big data analytics, Internet of Everything, and cloud computing) are used to alter radically:

i.   The way in which any government creates value;
ii. The manner in which it interacts with its populace and business partners, and
iii. How it competes in established and emerging markets.

Digital transformation is the key managerial imperative for today's business and institutional leaders wishing to recreate the way their institutions operate in a world of digital ubiquity. The key aspects of digital transformation are that it is disruptive in the introduction new ways of achieving results. Commonly digital transformation will cut through traditional established thinking and way of doing things –in particular the bureaucracy so often seen in the public sector.

For digital transformation to be successful, therefore it needs to be agile, flexible and deliver results fast. For example, the way Uber has disrupted the traditional taxi industry with a new way of approaching the problem of commuting service using network to connect, disrupting the old ways, was reusable connectivity, service disruption, service reusability or able to be replicated quickly and yet, service simplification with putting the power in the hands of the consumer.

The table below summarised the core characteristics of digital transformation.

S/N	Digital Transformation	Description
1	Inevitable	Total disruption of old business models
2	Irreversible	More efficient operations model
3	Tremendously Fast	Connectivity, Collaboration and Sharing
4	Uncertain in Execution	Future trends and application is unclear

Figure 9: Digital Transformation Characteristics

Defining digital transformation is to focus on the realignment of technology and business models to more effectively engage digital customers at every touchpoint in the customer experience lifecycle. However, digital transformation has different meanings to different people, but the most important thing to note is that investing in digital technologies such as social, mobile, big data, cloud, etc. does not have the exact meaning of digital transformation. It is about wiring individual technology efforts around a common vision which are supported by an updated and integrated infrastructure which aims to compete effectively as a unified business in connected markets. Summarily, with the most significant digital transformation initiatives being pursued and discussed, the following steps have been determined to improve the transformation approach:

- Improving processes which expedite changes to digital properties such as website updates, new mobile or social platforms;
- Updating website and e-commerce programs for a mobile world;
- Integrating social, mobile, web, e-commerce, service efforts and investments to deliver an integrated and frictionless customer experience;
- Updating customer-facing technology systems;
- Further research into customer digital touch points and
- Overhauling customer service to meet the expectations of digital customers.

For the government, the key to digital transformation remains to adopt technology as an enabler for something bigger – delivering on the common good.

DEFINING DIGITAL TRANSFORMATION OF PUBLIC SERVICES

Digital transformation, in short, enables the government to use digital tools and systems to provide better public services to its citizens and businesses. It also means a more efficient way of collecting revenue, i.e., taxes, fines, and so on. (*For instance, the government of Tanzania is meeting this objective through its M-Pesa – a mobile payment platform used for revenue collection*)[47]. However, digital transformation encompasses a far wider concept than just adding technology to government offices. It is not merely adding broadband Internet or digital service portals or creating dislocated websites as usually the case in developing countries. Rather, it is about the transformation of every single aspect of government operations, based on technologies which are already imminent and have become a lifestyle for many people, as well as the new ones which have a disruptive potential, such as cloud computing and big data analytics in proactive navigation and interactions.

Hence, digital transformation is much more about change, process re-engineering, disruption, integration and sharing than about technology – change in the attitudes and skills of public sector officials, change of business processes and models, and of making public services user-friendly, citizens-centric and accessible anytime, anywhere. Public services will drive the digital economy of the country, with the digital economy being defined as the global network of economic and social activities which are enabled by information and communications technologies such as the internet, mobile and sensor networks.

For Public Sector this means results can be achieved with a fresh approach including simplification, reusability but across wider sections of the public sector and serving the public by changing the control points and creating transparency.

HOW COUNTRIES SCORED ACROSS FOUR FACTORS ON THE
DIGITAL EVOLUTION INDEX (OUT OF 100)

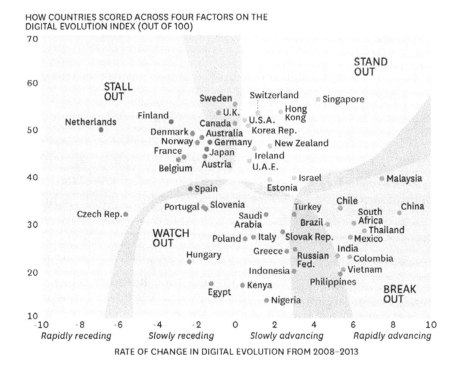

Figure 10: DEI of countries from 2003 - 2013 [48]

The Legend Explained:

- *Stand out countries are those that have shown high levels of digital development in the past and continues to remain on an upward trajectory.*
- *Stall out countries are those that have achieved a high degree of evolution in the past but are losing momentum and risk falling behind.*
- *Break out countries are those who have the potential to develop strong digital economies even though their overall score is still low, they are moving upward and are poised to become stand out countries in the future.*
- *Watch out countries are those that face significant opportunities and challenges, with low scores on both the current level and upward motion of their DEI (digital economy index).*

WHY DIGITAL TRANSFORMATION?

Digitalization allows governments to operate with greater accessibility, transparency and efficiency, and it has a dramatic effect on economic growth. Countries at the most advanced stage of digitalisation drive up to a 35% increase in economic benefits leading to an additional GDP growth of over 2.5% annually according to recent IDC study report [47]. Digitalisation can also have a huge impact on access to essential services in developing economies like Nigeria. Digitalisation plays a notable role in expanding access to information, education, justice, sanitation, water, sustainability and healthcare.

Figure 11: Digital in Nigeria [15]

Referencing the 2010 Corruption Perceptions Index published by Transparency International, the digital government development index created by the United Nations Public Administration Network to gauge the provision of electronic services by governments, and adding the Inequality-Adjusted Education Index, measured by United Nations Development Programme (UNDP) to show the extent of public education, a critical government service. It is found that greater digitalisation led to improvement on three broad counts:

- Digitalisation enables a society to be more transparent, increasing public participation and the government's ability to disseminate information effectively. (*A 10-point increase in digitalization increases the Transparency International index by approximately 1.2 points*);
- Digitalisation raises service effectiveness by about 0.1 points, and stimulating digital government efficiency accelerates digitalisation;
- This, in turn, gives the population more insight into public policy and functions. This can lead to increased political participation and greater development of human rights.

Digitalisation has been shown to support better delivery of public education, healthcare and other government services. In developing countries, a 10-point increase in digitalization results in an average 0.2-point increase in the Inequality-Adjusted Education Index. Nigerians today are more aware of their rights, have better access to information on public services and consequently have higher expectations of service levels, especially as they become accustomed to private sector organisations providing customised digital solutions and other benefits. Furthermore, some countries including Nigeria have empowered their citizens with "*Right to Information*" legislation. Citizens and businesses are therefore expecting better individualised public solutions and services, efficient and effective service delivery, burden reduction, transparency and participation.

According to the International Data Corporation (IDC), Information Technology spending in Nigeria will top $5.6 billion in 2016 as MDAs' race toward disjointed computerisation programmes. At the same time, economic and budgetary pressures are forcing governments worldwide to be ever more efficient, reduce costs and be more competitive. These challenges, coupled with the financial crisis have created renewed momentum for the modernisation of public administration. To meet these ever evolving demands, new and creative ways have to be found that improve quality and provide customised digital solutions, while reducing costs [48].

There are three broad scopes of digital government [49], namely:

- Digital Administration: This is the application of technology in the processes of government or the public sector organisation.
- Digital Services: This is the use of the Internet, the web and mobile technologies to improve the delivery of government or public services.
- Digital Democracy: This is the application of technologies with a view to enhancing the greater involvement of citizens in the decision-making processes of democratic institutions in the country.

Digital governance can be seen as the optimal use of electronic channels of communication and engagement to improve citizen satisfaction in service delivery, enhance economic competitiveness, forge new levels of interactions and trust, and increase the productivity of public services. A digital transformation encompasses the full range of digitalisation—from the core digitalisation of public services to digital infrastructure, governance and processes, including both front and back-office support services.

The public service needs to meet citizens' expectations continually rather than expect citizens to meet the Civil Service's bureaucratic and often tedious manual processes. It needs to become citizen centred in its approaches, its style, how it communicates and how it enables service user interaction [50]. The readiness for digital services acceptance in Nigeria is quite positive and encouraging as can be seen in figure 12 below – Nigeria Internet Adoption Rate as sourced from the World Bank Open Data Initiative portal.

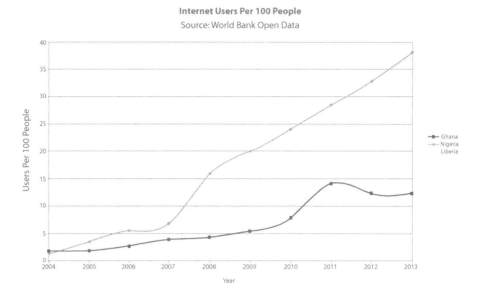

Figure 12: Nigeria Internet Adoption Rate

Once digital services are functional, Nigerians should be able to complete their transactions with the government easily in a digital environment. Their most common transactions with government should be entirely completed in a digital environment without undue interruption or repetition – take the case of biometric fingerprinting which is currently collected at multiple points in the country - *driver's license with Federal Road Safety Corps (FRSC), international passport with Nigeria Immigration Service (NIS), National Identity Card with National Identity Management Commission (NIMC), voters registration with Independent National Electoral Commission (INEC), Bank Verification Number with Central Bank of Nigeria (CBN), etc. leading to gross waste of public finances and citizens time.*

To eliminate such a situation, every domain of government should be able to deploy, integrate and use digital technologies in a manner that can increase the service-level standards, improve interactions with citizens, raise efficiency and save scarce resources through shared services. Digital technology can provide a foundation for innovative, integrated public services that cuts across organisational or MDAs boundaries while delivering to those in most needs, as well as offering services for business that promote growth and efficiency. A "*digital first*" approach will mean that the public sector will deliver online all services that can be offered online. And that citizen will be able to access public information and services in the same seamless and effortless way that they access services from the highest rated online commercial offerings. Service users will, therefore, use digital channels because they meet their needs not make-belief as currently seen by numerous government websites that lead no-where.

There are of course services [51] such as sensitive advice and counselling, where personal, face-to-face interaction will continue to be necessary and appropriate. Also, assistance must be provided in using digital channels where that is required, taking into account that many Nigerians will be unfamiliar with the new technology and that their level of literacy and language skills will vary.

Within this overarching definition of digitalisation of public service delivery, public services will also be

- accessible through a single, though not exclusive, the point of entry to all government services to help navigate through the broad public sector landscape;
- available with assisted access to take into account the differing capacities of users, including telephone or face-to-face support;
- secure, reliable, resilient, high quality and high performing integrated technology platform;
- made-up of systems of assuring identity that is secure which gives access to all public services;
- shaped by the needs of users and involve service users directly in their design;
- able to use technologies to enable personalisation of services and self-management;
- designed to take account of the methods and capabilities already provided by the private sector that are used extensively and intuitively by citizens, and

- joined up through the use of standard technology applications, frameworks and methodologies.

The following points are integral to implementing this transformation and are inherent to the prescribed processes while running concurrently as other public reform programmes or agendas:

- Strengthening accountability.
- Capacity development where necessary.
- Transformation of performance management and career development for all public servants.
- Integration of policy reform and implementation teamwork.
- Improvement of Data and Management Information System (MIS) to support government decisions.
- Achievement of structural reduction in public expenditure.
- Modernisation of government structures and functions.
- Greater recognition for the work of public servants.
- Promoting a culture of transparency and results.

DIGITAL TRANSFORMATION ESSENTIALS

To achieve the above-stated imperatives of digital transformation and to provide useful digitally enabled public services, strategic reform is required within the public sector, requiring the following four fundamental action points:

1. The re-organisation of existing administrative structures;
2. The rationalisation of existing MDAs and optimisation of government administrative procedures,
3. The optimum utilisation of human resources and
4. The strengthening of transparency and accountability within all MDAs.

Everyone has grown accustomed to simple and intuitive applications that collectively help us manage all facets and spheres of our lives, with the current rise in the use of digital devices. However, the government has struggled to keep pace with the increase in expectations across all industries. Digital innovation now cries to be at the heart of every enterprise (private or public) and to do things better, faster and cheaper, there arises a need to digitise internal government operations.

An important aspect of what makes major cities thrive is infrastructure, along with the analytics programs that allow people to use that infrastructure more efficiently. What makes a happy municipal citizens? Things like knowly exactly when the train will come, being able to go online instead of standing in line to access a government service, and being able to provide real-time feedback that informs[82]

Case Study: Oil-funded digital infrastructure

Azerbaijan's economy bears similarities with Nigeria - they are both oil-rich emerging markets, with an urban-rural digital divide. Oil comprises 75% of Azerbaijan's government revenues and 95% of exports and mobile penetration, like Nigeria, is split along urban and rural lines. Roughly 54% of the population is urban, with a mobile penetration rate of 141% versus the 46% of rural dwellers, who have a 69% mobile penetration rate. Using oil revenues, the Azerbaijan government made sizeable investments in the digital network. Coverage is high, at around 99% for mobile network and 61% for internet. Two state-linked companies, Delta Telecom and Azertelecom, using funds from the State Oil Fund of Azerbaijan (SOFAZ), operate the country's fibreoptic network, while two government-owned internet service providers, Bakinter.NET and Aztelekom have the largest market shares.

Specific digital infrastructure investments include a $38 million 2013 fibre optic network expansion and $130 million broadband expansion led by Aztelekom, in conjunction with SOFAZ. Indeed, SOFAZ transfers to the national budget increased by 93.7% between 2007 and 2014. These expansions support official targets; the national regulator wants 85% internet penetration by 2017 and a higher quality broadband provision of 10 Mb/second in all areas. However, the declining price of oil has stalled the government's ability to expand broadband internet access further, as SOFAZ funds have now been diverted from digital infrastructure expansion to budget consolidation. The oil price collapse shows that funding infrastructure from oil revenues can bear fruits only when prices are high, but stall when prices fall. Blended funding models, perhaps with donor assistance, may prove more sustainable in oil dependent economies.

A second lesson from the Azerbaijan experience is quality of service challenges. In Azerbaijan's rural regions, there is a high degree of infrastructure sharing, where companies share existing assets like base transceiver stations. This increases coverage, but service quality lags.

Low RoI coupled with bandwidth access fees mean that while coverage is good, companies have limited profit margins to invest in improved quality of service [81].

"If at first, the idea is not absurd,
then there will be no hope for it"
— Albert Einstein

Chapter 5

BLOCKCHAIN AND THE NEW DIGITAL ECONOMY

Today, bitcoin adoption has been influenced by different factors that have contributed to significantly make it the most volatile currency in the world. Despite the above, over 100,000 transactions are happening per day, and the volume is going north due to permission-less online innovation powered by bitcoin's underlying technology – blockchain. In November 2008, a paper was posted on the Internet under the name Satoshi Nakamoto titled *"Bitcoin: A peer-to-peer Electronic Cash System. This paper detailed methods of using a peer-to-peer network to generate what was described as "a system for electronic transactions without trust"* [49].

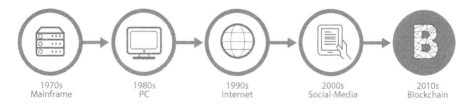

| 1970s | 1980s | 1990s | 2000s | 2010s |
| Mainframe | PC | Internet | Social-Media | Blockchain |

Figure 13: Blockchain - foundation of new digital economy [50]

Blockchain technology allows peers to exchange money directly without the need for a traditional financial intermediary, lowering the cost and increasing the speed of transactions. However, it is proving to be much more than a way to transfer monetary value. At its core, blockchain technology is a way to transfer any information in a fast, tracked, and secure way. As mentioned in preceding chapters, the idea of using blockchain technology to provide services traditionally provided by governments in a decentralised, cheaper, more efficient and personalised manner is just emerging. Many new and different kinds of governance models and digital services are possible using the Blockchain technology. Blockchain governance model takes advantage of public record-keeping features of technology – *the blockchain as a universal, permanent, continuous, consensus-driven, publicly auditable, redundant, record-keeping repository* [50].

Blockchain technology was created as a way to transfer value, specifically within the context of the digital currency Bitcoin — money in the form of data streams. As such, primary uses of blockchain technology include payments and other financial transactions. Blockchain enables peer-to-peer transactions by removing the need for a trusted intermediary verifying the transactions — a role that is required when peers do not know or trust each other, currently played by government regulators. Blockchain makes this possible by being a decentralised public ledger. This public ledger is not so different from the ledger that traditional financial institutions maintain, with a record of who owns what. The difference is that there is no bank or other single third party keeping the ledger and verifying the transaction — no one entity controls the ledger. Instead, the network, as a whole, confirms the transactions through a decentralised "*consensus mechanism*" as described in the section on blockchain transaction.

THE DEVELOPMENT OF BLOCKCHAIN TECHNOLOGY

While blockchain was originally developed as part of digital currency, people realise that at its core, it is a way to transfer any information in a fast and private way and that it can be useful for any form of information or value transfer that typically involves an intermediary. This realisation has spurred intense development activity in the market. In fact, people in the field are comparing it to the early stages of the development of the Internet, and there are similar levels of capital investment in start-ups related to blockchain technology services and applications as there was in the development of the Internet in the mid-1990s. Just as the Internet relies on services such as browsers and email clients to help consumers access its capabilities, blockchain technology's utility and continued development will rely on innovation by new service providers. Since the blockchain mechanism was originally conceived as a financial exchange tool for bitcoin, much of the innovation activity so far has been in financial applications. It is important to note, however, that a coin on a blockchain could easily represent more than bitcoin or money. It could represent a house, a car, stock, or even a vote or an identity. Arguably, a coin could represent any information or any piece of data. It is this realisation that is sparking growth in this sector, including the development of new applications and increased interest in this technology by governments. In summary, emerging blockchain development will enable:

- Creation and real-time movement of digital assets.
- Embedding trust rules inside transactions and online interactions.
- Time-stamping, rights and ownership proofs.
- Identity ownership and representation.
- Resistance to single points of failure or censorship.
- Creation of crypto-currency markets.
- Self-execution of business logic with self-enforcement.
- Running decentralised services.
- Selective transparency and privacy.

The blockchain revolution is so fascinating because it could be TWO completely different revolutions happening together both profound in their implications:

1. *Industry-level systems of record are providing massive efficiency gain for incumbents.*
2. *Censorship-resistant digital cash is providing a new platform for open, permission-less innovation.*

BLOCKCHAIN TRANSACTION

This technology is called "*blockchain*" because individual transactions are grouped into what is known as a block of data sets. Members of the public peer network seek to verify the block set, and each of these verifiers in the network retains a copy of the ledger on their individual hard drives. Block sets then build upon each other, with the data in each block being irrevocably linked to the blocks before it — hence the word "chain." As such, each "coin" on the blockchain is a string of encrypted data that identifies every transaction that coin was ever involved in — a sort of historical record. Imagine holding an N100 note in your hand and not only seeing the physical cash but also a chain of information identifying every transaction that the note was ever involved in. Because of this linkage of blocks and transactions, it is prohibitively impractical and computationally tough to modify a block once created and confirmed. This helps to keep the system secure while eliminating the need for a trusted intermediary. Figure 14 below illustrates a blockchain transaction [52].

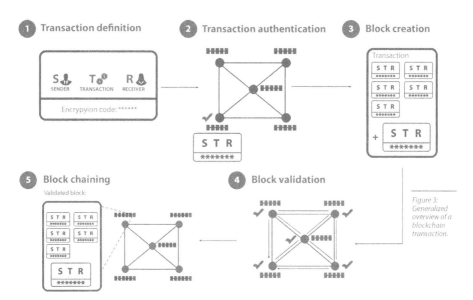

Figure 14: Blockchain Transaction [52]

A block chain is therefore a transaction database shared by all nodes participating in a system based on the Bitcoin protocol. A full copy of a currency's block chain contains every transaction ever executed in the currency. With this information, one can find out how much value belonged to each address at any point in time. Every block contains a hash of the previous block. This has the effect of creating a chain of blocks from the genesis block to the currently active block. Each block is guaranteed to come after the previous block chronologically because the previous block's hash would otherwise not be known. Each block is also computationally impractical to modify once it has been in the chain for a while because every block after it would also have to be regenerated. These properties are what make double-spending of bitcoin tough. The block chain is the main innovation of bitcoin – crypto-currency.

Honest generators only build onto a block (by referencing it in blocks they create) if it is the latest block in the longest valid chain – validator node. "Length" is calculated as total combined difficulty of that chain, not some blocks, though this distinction is only important in the context of a few potential attacks. A chain is valid if all of the blocks and transactions within it are valid, and only if it starts with the genesis block. For any block on the chain, there is only one path to the genesis block - coming from the genesis block. However, there can be several forks. One-block forks are created from time to time when two blocks are created just a few seconds apart. When that happens, generating nodes build onto whichever one of the blocks they received first. Whichever block ends up being included in the next block becomes part of the main chain because that chain is longer. More serious forks have occurred after fixing bugs that required backward-incompatible changes.

Blocks in shorter chains (or member nodes) are not used for anything. When the bitcoin client switches to another, longer chain, all valid transactions of the blocks inside the shorter chain are re-engaged to the pool of queued transactions and will be included in another block. The reward for the blocks on the shorter chain will not be present in the longest chain, so they will be practically lost, which is why a network-enforced 100-block maturation time for generations exists. These blocks on the shorter chains are often called "orphan" blocks. This is because the generation transactions do not have a parent block in the longest chain, so these generation transactions show up as an orphan in the list-transactions RPC call. Several pools have misinterpreted these messages and started calling their blocks "orphans". In reality, these blocks have a parent block, and might even have children [52]

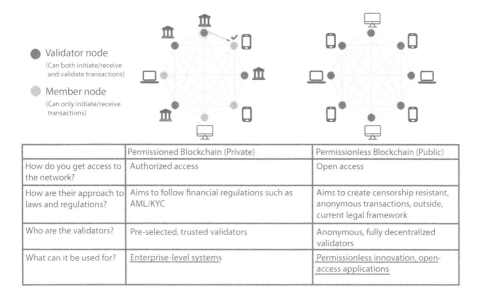

	Permissioned Blockchain (Private)	Permissionless Blockchain (Public)
How do you get access to the network?	Authorized access	Open access
How are their approach to laws and regulations?	Aims to follow financial regulations such as AML/KYC	Aims to create censorship resistant, anonymous transactions, outside, current legal framework
Who are the validators?	Pre-selected, trusted validators	Anonymous, fully decentralized validators
What can it be used for?	Enterprise-level systems	Permissionless innovation, open-access applications

Figure 15: Blockchain - validator and member nodes [50]

Although the blockchain technology that most people are familiar with is bitcoin, there are many different blockchains out there on the internet. At the highest level, these blockchains fall into two groups: public and private. In a public blockchain, the right to alter the ledger by participating in the consensus mechanism is open to anyone. Transactions are publicly available for anyone to read. Bitcoin and Ethereum are two prominent examples of the public blockchain. In a private blockchain, the right to alter the ledger by participating in the consensus mechanism is restricted to pre-selected individuals or institutions. Transactions may either be publicly available or limited to a select number of participants. Ripple is a prominent example, and private blockchain is also of growing interest for government and business use.

Figure 16: Strengths and Weaknesses of Blockchain Technology

Blockchain transactions are entirely different from our everyday transactions. They have unique attributes that offer users some potential benefits. These advantages are what have sparked the interest in this technology and innovation in the area of digital transformation for public services. On the other hand, as with any new technology, there are still many challenges associated with blockchain that is important to consider. The US Postal Service, Office of Inspector General, collaborated recently with Swiss Economics to outline the benefits and shortcomings of blockchain technology. These strengths and weaknesses emerged stated below are a result of this study [52].

BLOCKCHAIN STRENGTHS

Lower Cost of Transactions

Due to the decentralised nature of blockchains, adopters have the ability to make online transactions for a fraction of the fees charged by current intermediaries such as financial or legal institutions. Credit card companies charge a fee per transaction for processing, which is a cost that is usually borne by merchants but which can also be passed along to buyers through higher prices or an additional fee for purchasing with a credit card. Remittance service providers charge senders an average of 8% to transfer money to family overseas. In the financial services sector alone, Spanish bank Santander estimates that blockchain technology could save banks around the world $15-20 billion annually in the settlement, regulatory and cross-border payment costs. Outside of the financial services sector, IBM has suggested that blockchain can help reduce infrastructure and maintenance costs of scaling the Internet of Everything by allowing connected devices to "share computing resources without dependency on a central cloud or server, thereby optimising resource utilisation and cost." Other cost savings of the technology are only just beginning to be investigated [52].

Faster Transactions

Blockchain transactions are processed much more quickly than most traditional data transfer systems, usually in a matter of minutes. With blockchain, time is saved by the elimination of intermediary institutions such as clearinghouses that make sure banks or others parties have matching records [52].

This feature is especially significant when it comes to payments, which can take hours, days, or even weeks to process. For example, when trading stocks or bonds, it usually takes three days for a transaction to settle and for the participants to have their funds available.

This is true even for electronic transactions where the information exchange may be immediate, but it may take three days to receive payment. Real estate sales are also costly and time-intensive, often taking weeks to schedule a time for closing with thousands of dollars in closing costs. With smart property, selling a house could be as simple as transferring a coin. Other applications, such as not having to present yourself in-person to vote or notarize a document could save time and increase the convenience of these processes. Blockchain allows for faster, more efficient, and more customizable transactions [52].

Geographical Freedom of Transactions

Transactions across a blockchain are not bound to geographical limits. Given the virtual nature of the system, it does not matter whether an individual sends data to a neighbour or someone on the other side of the world. Also, as blockchain do not use intermediaries, which are bound by country-specific regulations, transactions can cross national borders with less friction. This makes blockchain well suited for international trade [52].

Irreversibility of Transactions

Blockchain-based payments are irreversible; once payment is issued, it can only be reversed by asking the receiver to pay the same amount back in another transaction. This feature is ideal for lowering transaction risk for a payment recipient, allowing merchants to be sure that buyers cannot cancel a payment after the sale of a good or service (the way they can with credit card purchases). This alleviates fraud risks and security payment costs for merchants. On the other hand, buyers may not view this as an advantage. This is because conventional card- and bank-based payment providers, acting on behalf of the purchaser can reverse transactions to protect customers against fraud, such as being overcharged or if a good is defective. However, the irreversibility feature is not only beneficial to merchants. It applies to other application areas as well; including the transfer of property where there would be no way, for example, for someone selling a house on a blockchain to reverse the transaction and get the deed back after receiving payment. This feature would also mean that records could not be tampered with, altered, or undone after they have been created, making blockchain a highly transparent and auditable records management tool.

Increased Privacy of Transactions

Currently, completing an e-commerce transaction or enacting a legally binding contract requires participants to disclose their personal information to another party, such as an e-commerce platform. Transferring information across a blockchain is similar to paying with cash: there is no need to disclose any personal information such as a person's name, address, credit history, or credit card number. Individuals only disclose their wallet information, which is an alphanumeric "address." In addition to protecting user privacy, blockchain transactions substantially reduce the risks of identity theft and fraud that are common to other forms of transaction or payment, such as credit cards [52].

BLOCKCHAIN WEAKNESSES

Technological Barriers

Blockchain technology is new and very different from most of the traditional technologies that people use. As such, in its current form, it requires above-average computer literacy to use properly, which acts as a barrier to entry for businesses and individuals that are interested in applications but do not know where to begin. This can limit access to the new technology for non-tech-savvy users, and can expose them to fraud risks. Furthermore, blockchain's decentralisation means that there is no central customer care resource if users need assistance.

Security Concerns

Although the Bitcoin blockchain has so far not been compromised, service providers (such as wallet providers or exchange services), are vulnerable to attacks. Furthermore, the privacy of transactions seen as a benefit to many is also a security concern. Not knowing the identity of the individual on the other side of the transaction makes it difficult to resolve issues that may arise and can place users at risk of fraud [52].

Limited Access

At present, access to blockchain applications is provided by online exchanges. Physical touch points, such as Bitcoin ATMs and other physical service locations are scarce and scattered. Service platforms are mostly new start-up firms with little reputation and lack physical exchange points.

Regulatory Uncertainty

A lot of progress has been made in recent years, but there is still no international — or even interstate — agreement about how to regulate blockchain applications. Current regulations focus on financial applications of blockchain technology. It remains to be seen how applications such as smart contracts, smart property, and records management will be regulated. Up to this point, some government entities have emphasised instituting consumer protections while letting innovation continue to develop, but others have imposed more restrictive regulations. For example, the state of New York requires a "*BitLicense*" for businesses operating in this space, causing many start-ups to leave the state. This regulatory uncertainty, coupled with speculation, has led to other problems, such as exchange rate volatility in the crypto-currency applications such as Bitcoin.

NEW BLOCKCHAIN APPLICATIONS AND SERVICES

"Blockchain technology continues to redefine not only how the exchange sector operates, but the global financial economy as a whole."
— *Bob Greifeld, CEO, Nasdaq*

Information Technology developers are beginning to create and market novel uses of the blockchain, which has the potential to disrupt any sector that uses intermediaries to verify or track the transfer of information. Some of the major application areas include financial services, the transfer of property, the execution of contracts, authentication and identity services, network and device management, and records management. This has led the Institute of Electrical and Electronics Engineers (IEEE) to suggest that "*the possibilities are endless, and that money is only the first, and perhaps the most boring application enabled by blockchain technology.*" Ethereum is a decentralised platform that runs smart contracts: applications that run exactly as programmed without any possibility of downtime, censorship, fraud or third party interference. These apps run on a custom-built blockchain, an enormously powerful shared global infrastructure that can move value around and represent the ownership of property. This enables developers to create markets, store registries of debts or promises, move funds in accordance with instructions given long in the past (like a will or a futures contract) and many other things that have not been invented yet, all without a middle man or counterparty risk. The project was crowd-funded during August 2014 by fans all around the world. It is developed by the Ethereum Foundation, a Swiss non-profit, with contributions from great minds across the world [53]. The question now is when and how will Nigerians use Ethereum to create a new digital economy. Table 1 describes common application areas pursued by many of the start-ups in the blockchain technology space.

Table 1: Emerging Blockchain Application and Services [52]

S/N	Type of Blockchain Service	Intermediaries Replaced	Sample Companies in the market
1	Financial Services: Services involving the exchange of money that can now be performed on the blockchain. Examples: Money transfer, Remittances & Payment, Stock Exchange, etc.	Banks, Credit card companies, clearinghouses, and other intermediaries in the financial industry	Bitpay, Coinbase, Ktaken, BitPesa, Bitcoin Venezuela, Counterpart, Bitso, Bitt, etc.
2	Smart Contracts: Verification and Enforcement of agreements negotiated between two or more parties, such as escrow services or wills, is written into the blockchain	Lawyers, sharing economy platforms (such as AirBnB or Uber)	SmartContract, Codius, new System Technologies, Bitnpaky, UbiMS, BitHalo, Lighthouse
3	Smart Property: Tracking ownership of physical and non-physical property, such as cars, real estate, stocks, and other assets through a blockchain	Lawyers, stock markets, real estate and other sales agents	Symbiont, Mirror Labs, Secure, Assets, Bitshares, equitybits, SXMarkets, UNA, Everledger
4	Authentication Services: use of information in addition to a user's public wallet address, such as a home address or social media profile, to verify and authenticate a user's identity and activity on a blockchain. Services include identification, verification and proof of ownership	Government agencies, such as motor departments or the Social Security Administration, and other identity brokers	Onename, bitnation, BlockCDN, Colu, MyPowers, TRSTim, Blochai, Bitproof, Proof of Existence, etc.
5	Records Management: The maintenance of official records and registers in a transparent, secure, and auditable manner by using a blockchain. Includes the management of health records, government records, and voting	Hospitals that manage patient, records, governments	BitHealth, Agora, BitCongress, Follow My Vote, BTC Blockchain Apparatus
6	Network & Device Management: Allowing connected objects to communicate with each other directly over a blockchain. This enables decentralised Internet of Everything management and network storage	Centralised cloud computing, human intervention required to manage devices	Chimera IoT, Storj, Filament, TilePay, etc.

BLOCKCHAIN ADOPTION AND FUTURE OUTLOOK

In addition to an increasing array of diverse applications being developed by start-up companies – as shown above, there is an increased use of blockchains by major institutional players especially governments around the world. In January 2014, Overstock.com, a US based online retailer became an early adopter of blockchain technology by becoming the first major company to enable payment in bitcoin on its website. It has since received approval to issue stock through the Bitcoin blockchain as well. Since then many companies have announced plans to use blockchain, invest in blockchain start-ups, or at the very least research how the technology could affect their business or more recently their future.

As one would expect, some of the interested companies are in the banking and financial services sector. UBS AG (a Swiss global financial services company) has opened a research laboratory to explore the application of blockchain technology to their financial services business. Citibank group is testing different uses of blockchain in their laboratory and has gone so far as to experiment with their digital currency: **CitiCoin**. This, in itself, is not very difficult – anyone can create a crypto-currency in a few seconds with a bit of programming knowledge. JPMorgan Chase, America's largest bank has just begun to investigate how blockchain could be used, for example, to deal with data errors in the bank's loan funds. Additionally, the stock exchange, Nasdaq revealed that it would soon record trades in privately held companies on a blockchain-based system.

Outside the financial services sector, some additional blockchain applications are beginning to gain traction — particularly with governments. In the records management space, the United Kingdom Government is starting to research a role for blockchain in keeping registers. Erstwhile Russia has been unfriendly towards the bitcoin, it is now showing enthusiasm towards its underlying blockchain technology. The Russia Central Bank moved recently to embrace the blockchain technology, as stated by Andrei Lugovoi, Deputy Chairman of the Russian State Duma Committee for Security and Anti-Corruption which further generated hope for the future of crypto-currencies. The Liberal Alliance in Denmark was the first major political party in the world to vote using the blockchain technology, in an internal election, and similar systems were later used in Norway and Spain.

Citizens of Estonia will soon be able to manage their health records and will have access to them via a blockchain-based database. In the area of identity management, the Estonian Government has announced that residents with an "*e-Resident*" digital identity will be able to use blockchains to notarize marriages, births, contracts, and more. IBM is building a coin-less blockchain for smart contracts and hoping it will help move the technology into the mainstream. Finally, Australia Post recently announced that it is exploring the role of blockchain technology as a tool for digital identity management.

To promote innovation and adopt next generation technologies at a global level, Dubai Museum of the Future Foundation recently announced the launch of Global Blockchain Council. Al-Aleeli, CEO, Dubai Museum of the Future Foundation said that "*the significant growth in the volume of transactions using Blockchain platforms during 2015, which reached 56%, refers to the great opportunities that can be utilised through the optimal application of this technology in the relevant sectors.*" According to him, the global investments in blockchain could reach $300 billion over the next four years. The Global Blockchain Council will explore and promote blockchains and digital currencies as well as study its advantages and disadvantages while working on ways to utilise the technology in the best possible way.

A new draft repor by the European Parliament on virtual currencies stresses that virtual currencies and blockchain technology can contribute significantly towards consumer welfare and economic development by dramatically lowering transaction costs for payments and transfer of funds while enhancing the speed and resilience of payment systems, and allowing transactions to be tracked in case of a malfeasance. The report calls for the creation of a horizontal

Task Force DLT (distributed ledger technology) under the leadership of the Commission to facilitate the necessary technical and regulatory expertise to support the relevant actors (at both EU and Member State level) in efforts to ensure a timely and well-informed response to the new opportunities and challenges. Meanwhile, the European Central Bank (ECB) showed "*openness to new technologies*" and said that it intends to assess the relevance of blockchain and distributed ledger technology for various banking services like payments, securities settlement as well as collateral [53].

The developments around bitcoin and the blockchain technology are fast evolving. Soon we will be able to get a "*proof for everything*" – a vision of online trust checking similar to Google searches. We can already imagine a wide variety of blockchain based services that will become the new intermediary "trust authorities" for all transactions whether public or private. The emerging promise of blockchain technology is captured as follows:

1. Proof of identity: - get your identity verified by a blockchain-based certificate authority, supported by national public key infrastructure online.
2. Evidence of existence: - record audio or video message or even take a photo or share a file with existential proof for all to believe.
3. Oracles as a business: - Oracles will become reference authority, because of their inherent ability to contain useful information that is always updated.
4. Proof of Affidavit or Ownership: - get sworn declaration of facts or verify ownership or purchase of anything from arms to lotto ticket, land, etc.
5. Smart Contract as a service: - consult smart contract directories directly from your internet browser.
6. Proof of Location: - prove to your partners that you were in a given location with supporting timestamp.

Government and governance related applications that are ripe for blockchain technology adoption in the context of digital transformation in Nigerian public service are as follows:

- Marriage registration
- Procurement and auctions
- Passport issuance
- Benefits collection
- Land Registration
- Licenses
- Birth and Death Certificate
- Property Ownership
- Motor Vehicle Registration
- Taxes
- Voting
- Government Bonds
- Fillings and Compliance Monitoring.

Case Study: Bitland is running a Blockchain Land Registry in Ghana

Due to the tamper-resistant nature of distributed ledgers, the project hopes to give the African citizens a better method to survey land and record title deeds using the Bitland Blockchain. It is engaging with about 28 communities in Kumasi, Ghana and provides immutable records of title ownership to those who normally would have none.

Bitland is an initiative to provide a Blockchain based infrastructure to give disenfranchised local communities access to register their land titles to avoid land ownership disputes. As well, the project is design such that a person who registers a property through Bitland is granted automatic access to be able to source micro-loan. The newly registered land acts as collateral for the loan thereby opening up opportunities for new investment and development.

Research studies and observation have continually shown that part of the land tenure problem in Ghana is associated with corrupt government Officials or community Chiefs who engage in record manipulation or unjustly distributing land, having a public ledger becomes an objective way to keep open track of land title changes. The Blockchain entries are immutable, it is not possible for any official or individual to change record transactions in favour of any party. This is working to help stamp out rampant corruption in the land registry system in Ghana.

Bitland Nigeria is currently working to open offices in Nigeria with similar preposition to land title owners and government [51]. Bitland Business Development Centre Nigeria Limited Open shop in Lagos, August 2016.

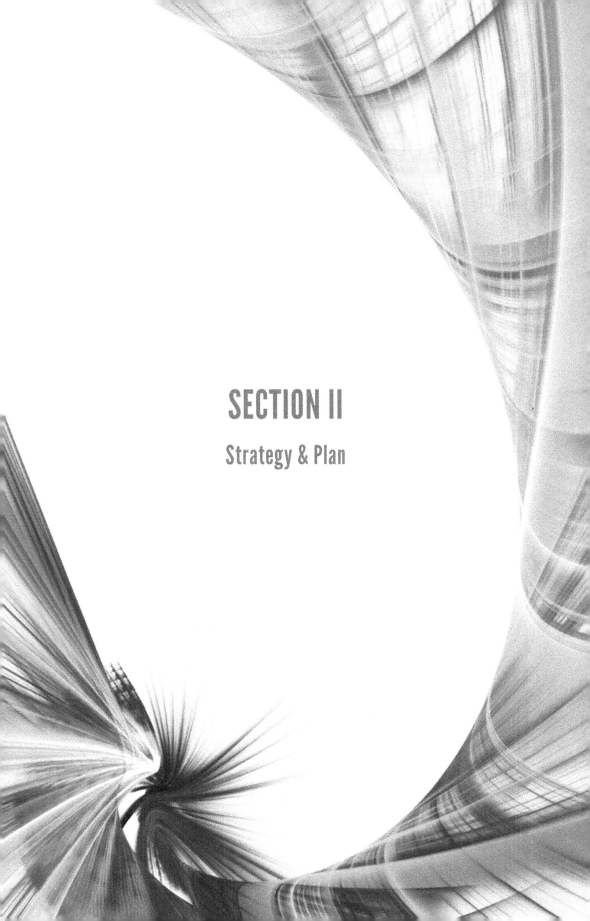

SECTION II

Strategy & Plan

"The only wrong move when it comes to Digital Trans-
formation is not make any move at all"
— *Dider Bonnet*

Chapter 6

DIGITAL TRANSFORMATION AND THE FUTURE OF GOVERNMENT

In the future, we can easily see an increase in the geographical spread of broadband connectivity and mobile device use. Computers will become faster, even smaller, and more intuitive than ever. Life will start changing at a more rapid rate. Technology will squeeze out old working approaches through new business models, but it will also create new jobs and ways to work. The evolution of digital technologies will significantly impact people, businesses and governments, but instead of shrinking the operating fields as predicted by some, digitalization will create more opportunities than ever seen before.

Gartner estimates that by 2030, 75% of businesses will be digital, or have digital business transformations underway. However, only 30% of those efforts will be successful due to lack of talent and technical expertise [91]

Henry Ford was once attributed as saying, "*If I had asked people what they wanted, they would have said faster horses.*" Technology-centric goals, Internet of Everything and big data are important, but they will only take digital transformation so far. The truly life-altering changes often come from visionaries who can picture the future and the steps that will bridge the gap between current and future digital technologies. The people, businesses and governments that succeed in the next 10-20 years will be those that recognise the need to change their strategic mind-sets and outlook, embrace constant change, take calculated risks and are completely "rubber flexible" with technology offering.

Adopting a longer-term perspective is a critical factor. One must look beyond 2020 now and ask '*what is this industry like?*' It's more than asking '*what it's going to look like*', but '*what should it look like?*' It's not something that we necessarily do very much because most government entities are so focused on operations and literal survival as they are getting through quarter to quarter." – according to Glen Hiemstra (founder and owner of Futurist.com), speaking on the future of digital transformation. Managing Director, Africa at Avaya, Hatem Hariri, has said digital transformation is key to Nigeria's and other African countries' future. He said governments and enterprises around the globe are looking at digitalization strategies to drive operational excellence, greater competitive differentiation, customer and citizen satisfaction, and providing them with a more connected experience. With the majority of Africa's population under 30 years, and the World Bank Group predicting that the continent's population will reach 2.8 billion by 2060 – more than a quarter of all people on earth – digital transformation remained key to its successful future, he said.

"*Governments want to be more accessible to their citizens – that means communications, which is where we help deliver smart government, making life easier for citizens by working in partnership with government organisations to transform everything to digital. Client-tailored and outcome-focused digital and smart services elevate organisations of every scale and accelerate growth through their digital journeys.*

"While Africa's predominantly young, tech-savvy, population means that the awareness of the so-called Third Platform technologies – Big data and analytics, cloud, social and mobile – is high, adoption is still uneven, given the diverse nature of the market. However, it very much depends on the country.

"Some countries have good bandwidth, good infrastructure and good education, and are therefore very advanced in their adoption of new technology. If you look at countries like South Africa, Kenya, Morocco, and Algeria, their cloud value proposition is gaining tremendous traction, how they integrate video as a service, unified communications and so on. Many countries here have very advanced technology solutions that are ahead of other regions – for instance, we have seen the use of mobile phones to conduct banking transactions being pioneered in sub-Saharan Africa," Hariri who spoke in Lagos, said.

He said other countries may seem behind in technology adoption but they are catching up fast as the infrastructure develops, stressing that on a positive note, African countries have all opened the market for service providers to come in, so each country has several service providers competing to provide better service and a better customer experience.

"Africa is fertile a land, whatever you put into it will bear fruit. We are investing more and more into the region, we are increasingly looking at other countries we can invest in because of the potential here. Africa is the hidden jewel of the world that is set to become the shining jewel. We are in Africa to stay and will participate in and contribute to the continent's growth," he said.

"While Avaya has had a presence in Africa since its inception, like other companies, it made the decision to step up its investment in the region after the global recession as part of a greater focus on global growth markets. Having established its first office in Kenya, today Avaya has offices in Nigeria, South Africa, and Egypt, with a presence in Morocco and Algeria, as well as legal entities in Zambia, Tanzania and Ghana

– which can be turned into offices at any time; through its industry ecosystem of partners, Avaya has a presence in every African country today," Hariri said.

Any vendor that comes to Africa is expected to understand Africa because the problem most companies have is that they think of Africa as one country, where one approach will work, not 50-plus countries, each with their own political, educational, and technological systems. *"Africa has more than one billion people, with a huge mix of cultures, languages and development levels; some companies want to work in Africa remotely, but you can't do that in Africa – you need to have a local presence. There's an old adage that says 'people buy from people' – but it is definitely the case in Africa. If you don't have a presence on the ground, you can't really understand the diverse cultures and what is required to deliver effectively."*

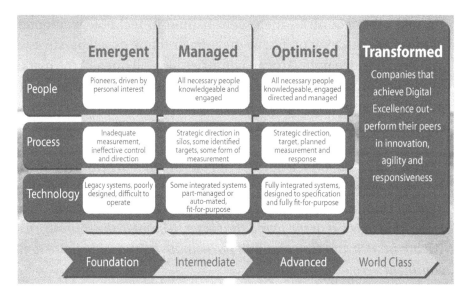

Figure 17: Roadmap to Digital Transformation [92]

On-going research in the field of digital transformation continues to show that success is not all about technology and innovation – in fact, the 2015 Digital Business Global Executive Study and Research Project sponsored by MIT Sloan Management Review and Deloitte identifies strategy as the key driver [93]. The ability to digitally reimagine the public services is determined to a large extent by a clear digital strategy supported by a strong leader who fosters a culture able to change and invent the new. While these insights are consistent with prior technology evolutions, what is unique to digital transformation is that risk taking is becoming a cultural norm as more digitally advanced countries seek new levels of competitive advantage.

STEP 1: Understand where to start the process.

STEP 2: Define excellence for the digital transformation effort and the new business model.

STEP 3: Map the exercise against capability and competency frameworks.

STEP 4: Understand that skills alone don't achieve digital excellence.

STEP 5: Ensure support staff have the skills to recruit and manage change.

STEP 6: Ensure senior people in government understand the need for change to a digital environment.

STEP 7: Digital marketing is digital business as well.

Michael Wade, IMD's professor of innovation and strategy suggests three core capabilities to successfully navigate the vagaries of digital transformation and protect against digital disruption as it's so often the case these days:

i. Hyperawareness - In the public sector, for example, the Port of Hamburg, Germany, has become more hyperaware by using sensors to enable better decision-making throughout its operations. People receive data at the right time so they can invoke the proper processes when needed. An integrated traffic management system (waterways, roads, and rail) allows the port to manage bridge closures and roadway congestion that tends to increase when ships are offloading. By integrating its traffic management, the port cut network costs by 78%, reduced standstill times for trucks and cars on the port grounds, lowered carbon dioxide emissions, and enabled cargo carriers to work more efficiently [94].

ii. Informed decision-making - The City of Chicago benefits directly from an ecosystem of outside application developers who use city data to develop new services and applications. For example, a private developer created SweepAround.us, a private app built on city data that alerts citizens when street cleaners are coming so they can move their cars to avoid tickets or towing. Another app uses crime and accident data to determine the safest routes by which to walk to school. City officials admit they do not have the resources to analyse this data and develop user-friendly apps themselves, but they see them as a helpful extension of the services and information that the city already provides [95].

iii. Fast execution - The City of Amsterdam's Smart City strategy typifies another digital transformation approach. It includes 47 Internet of Everything projects, such as smart energy grid systems, street lighting, parking applications, building management, and public Wi-Fi. Many of these projects span multiple city departments and involve private sector stakeholders. At the centre of Amsterdam's digital strategy is an open Information Technology infrastructure that will provide a platform for Internet-based innovations for years to come. With this infrastructure in place, Amsterdam can develop and test multiple new services rapidly.

Digital disruption threatens organisations of all kinds, whether public sector or private, but it also leads to the creation of new value and opportunities. By its very definition, disruption overturns entrenched habits and ways of thinking—and disruptive innovation is both creative and destructive. Smart businesses and governments alike, are learning to innovate and compete at the fast pace of the Internet's next wave - the Internet of Everything.

Big data and the Internet of Everything both represent a global market trend driven by a surge in connections among people, process, government, data, and things—from 15 billion today to 50 billion in the next decade. As these connections multiply, the result is exponential change, creating new revenue streams, better customer and citizen experiences, and new operational models that deliver ever-greater efficiency and value. More than perhaps any technological advance since the dawn of the Internet, however, Internet of Everything holds tremendous potential for helping public sector leaders transform the citizen experience. Already, some forward-thinking government institutions (federal, state, and local governments), healthcare organisations, educational institutions, utility companies, and non-governmental organisations (NGOs)—are already seizing the opportunity. They are using Internet-enabled devices and solutions to increase efficiency, reduce costs, save energy, and, most importantly, improve the lives of citizens. Nigerians cannot afford to be left behind.

DIGITAL TRANSFORMATION AND BLOCKCHAIN TECHNOLOGIES

The Blockchain technology is emerging at a level perhaps higher than the World Wide Web in significance. Its significance and promise in the area of digital transformation are huge especially as it is supposed to be more decentralised, more open, more secure, more private, more equitable and more accessible. Interestingly these are the very same promise of digital transformation blockchain applications are increasingly taking a shot at replacing legacy web applications and becoming the preferred model for online transactions.

In the book "*The Business Blockchain: Promise, Practice, and Application of the Next Internet Technology*" by Mougayar William, blockchain is seen as the next revolution in Internet technology. It is said to be a marching phenomenon, slowly advancing like a tsunami, and gradually enveloping everything along its way by force. Blockchain technologies are also considered as enormous catalysts for change that hit at governance, ways of life, traditional business models, society, and global institutions directly. The promise of this technology is its ability to loosen up trust, which hitherto has been in the hands of central institutions and allows it to evade old control measures and guarantees a freer, open and more trusting society.

The blockchain technology is filled with a lot of potentials, one of which is to transform the international monetary ecosystem. Because we live in a world where to access financial services, you have to pay more even though you earn less - one thing which is very obvious is with money remittances.

Sending money from say a developed country to another, it definitely would cost some good cash but sending from sub-Saharan Africa requires an upwards payment of 15% of the remittance agency that transfers the money. Not just people are affected, businesses also. However, some economies are excluded from financial markets they include some banks which have chosen to withdraw because of fines and increasingly haphazard regulation. What is blockchain technology?

It can be described as a distributed payment database in which every member of such network has a copy of the same digital ledger and there is a regular update on the ledger as soon as a transaction is complete. A solid record of the payment is made whenever there is an exchange of money, causing each party to have an access to secure accurate reference data, thus making it difficult to manipulate any financial information.

For a proper definition, Blockchain technology is seen from three complementary aspects that are viewed mainly as technical, business and legal capabilities. Technically, Blockchain is a back-end database transactional system that maintains a distributed ledger that can be inspected openly. In terms of business, the Blockchain is an exchange network or peer-to-peer network for moving transactions, value and assets between users without the assistance or support of intermediaries. Lastly, Blockchain can legally validate transactions, replacing the need for a trusted entity.

Blockchain multiplicity of functions strongly supports emerging digital transformation efforts. Its under-listed properties are supportive of future digital services in government:

- Crypto-currency
- Common Computing Infrastructure
- Common Transaction Platform
- Decentralised database System
- Distributed Accounting Ledger
- Common Development Platform
- Open Source Software
- Financial Services Marketplace
- Peer-to-Peer Network
- Trustee Services

Case Study: Shared Infrastructure

Sharing infrastructure can be a highly effective means of enticing private operators to enter lower income regions where returns on investments are not sufficient to attract their participation. There are two main forms of supply-side infrastructure sharing methodologies available to regulators:

- Intra-sectorial: Telecommunications and Internet operators share assets (e.g. base transceiver stations, mast, power sources, etc.).
- Passive infrastructure: Building telecommunications networks (e.g. fibre optic cables) into new infrastructures such as buildings, roads and railways.

Intra-sectorial approaches has been pursued in Madagascar, which experimented with shared towers, in which a communications tower tender was given to a consortium of operators federated around an infrastructure company to which operators contributed funds and purchases services. Shared broadband backbone is a second approach, taken in Burundi, where mobile and internet operators formed a private company to operate the national backbone as a wholesaler, helped by a national subsidy to ensure national coverage.

*"Digital Transformation, therefore, requires redesign
and re-engineering on every level – people, process,
technology and governance."*
— Alan Brown

Chapter 7

PUBLIC SERVICE DIGITAL TRANSFORMATION

Improved public service delivery begins with knowing whether the services offered are working as intended. But far too often, public service providers lack the means to solicit citizens' feedback. When feedback is available, the data typically represents the interests of only a fraction of users. This is particularly the case in Nigeria where persistent underdevelopment of infrastructure, including roads, Internet access, and electricity, constrains the ability of the country's poor to provide feedback. Those who stand to gain the most from effective public services have the fewest opportunities to contribute to their design, hence the need for a comprehensive digital transformation exercise.

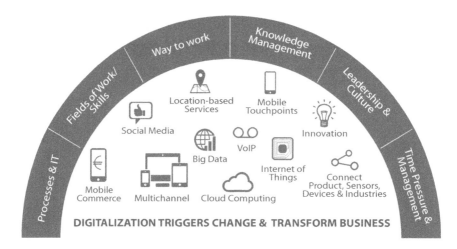

Figure 18: Digital Transformation [55]

Digital transformation in public service promises to establish a set of standards and practices for the design and use of various government digital services. It aims to be more focused on the needs of users, making the processes of applying for government documents like licenses, passport, company registration, etc. more efficient and easy to use. At the heart of this change are the users - people who need to use government services, from the trivial to the life-changing support services. All services, all platforms, everything to be built on this regime will focus on user benefits and experience.

DIGITAL TRANSFORMATION STRATEGIC APPROACH

Open Data – many individuals, organisations and government institutions collect a broad range of different types of data to perform their legitimate business. Government (Federal, State and Local Government) is particularly significant in this respect, not only because of the quantity and centrality of the data it collects, but also because most of that government data is public data by law, and therefore could be made open and made available for others to use. Opening government data to the public could immediately revolutionise the national economy, and the under-listed advantages are achievable:

- Lead to radical and remarkable transparency.
- Lead to incredible and unpredictable innovation where public data is shared and brought together in new ways.
- Lead to citizen empowerment, democratic accountability and good governance.

Open Standards – the use of open standards can help provide inter-operability and maximise access to resources and services across all MDAs. Open standard is also seen as standard adopted and maintained by a not-for-profit organisation, and its on-going development occurs by an open decision-making procedure available to all interested parties, especially government institutions. Adopting open standards as a strategic objective for digital transformation, delivers:

- Interoperability across devices and suppliers.
- Freedom from lock-in to any one vendor.
- Innovation from a level playing field for many companies, including cutting-edge start-ups.
- Access to resources is not dependent on a single application.
- Platform Independence: To ensure that access to resources is not restricted to particular hardware platforms.
- Quality scholarly resources can be preserved and accessed over an extended time frame.
- Architectural framework for technology developments is robust and can be further developed in the future.

Open Source Software – is software whose source code is available for modification, enhancement or change by anyone freely. The source code is the part of the software that most computer users don't ever get to see; it's the code computer programmers can manipulate to change how a piece of software works or behaves. Its adoption delivers the following advantages:

- The use of open source software radically reduces cost, improves delivery and enables innovation.
- Open source software delivers better control to programmers and most times is more secure and stable for educational use.
- Open source software is better at adhering to open standards than proprietary software.
- With closed source software, you have nothing but the vendor's claims telling you that they're keeping the software secure and adhering to standards, for example. It's a leap of faith. The visibility of the code behind open source software, however, means you can see for yourself and be confident.
- Open source software, is typically much less resource-intensive, meaning that you can run it well even on older hardware. It's up to you--not some vendor--to decide when it's time to upgrade.

Open Government – is the governing policy which holds citizens' right to access documents and proceedings of the government to allow for effective public oversight and visibility. In its broadest sense, it completely opposes reasons of state and other considerations, which have tended to legitimise extensive secrecy. Open government guarantees:

- Transparency. In a democratic society, citizens need to know what their government and public services are doing. To do that, they must be able to access government data and information freely **and to share that information with other citizens. Transparency isn't just about** access; it is also about sharing and reuse - often, to understand any material it needs to be analysed and visualised and this requires that the material is **open so that it can be freely used and reused.**
- Releasing social and commercial value. In today's digital world, data is a key resource for economic, social and commercial activities. Everything from finding services to building a sophisticated search engine requires access to a broad variety of data, much of which are created or held by the government. By opening up data, the government can help drive the creation

of innovative businesses and services that deliver social and commercial value in no small measure.

- Participatory Governance. Most of the time, citizens are only able to engage with their governance sporadically - maybe just at an election every four years. By opening up data, citizens are enabled to be much more directly informed and involved in decision-making. This is more than transparency: it's about making a full "read/write" society, not just about knowing what is happening in the process of governance but being able to contribute to it.

Cloud Computing – are computing services made available to people on demand through the Internet from a cloud computing platform. The idea behind this type of service is to provide fully managed, easy, and scalable access to applications, resources and services by a service provider. A cloud service can dynamically scale to meet the needs of its user population, and because the service provider supplies the hardware and software necessary for the service, there's no need for any particular government ministry, department and agency to provide or deploy its resources or allocate technical manpower to manage their individual services. The reasons to use cloud services are many and varied:

- Cost reduction – compared with private on-site hosting, the price of deploying applications in the cloud can be far less due to lower hardware costs from more effective and efficient use of physical resources.
- Universal access – cloud computing providers allows remote access to application systems and work through the internet.
- Flexibility – Cloud services allow people to switch applications easily and rapidly to use the solution that suits their needs best.
- Sustainability – the average energy requirement for computational work is far less in the cloud scenario than privately hosted deployment. Cloud computing provides a veritable room for resource sharing.

Open Procurement – government will spend huge amounts of money – some N2.24 trillion or 40% of 2016 budget on contracts to deliver various goods and services to Nigerians and its MDAs [60]. These procurements are tangible benefits, where our collective resources get converted into the roads, schools, hospitals and transportation that people care about. This is the crucial stage at which government's promises to its people either delivers or fails miserably. The tradition of continued failure in government contracts is now a clear and present danger to national development. Open contracting or procurement can help mitigate the risk of failure commonly associated with public contracts and public procurement. Transparent and participatory procurement can also cut costs dramatically, ensure better value for money, reduce fraud and corruption, provide more opportunities for businesses, and empower people to hold local, state and national government personnel to account.

Open contracting, the use of data, disclosure and engagement throughout the full procurement cycle, is an essential, but still a relatively new concept in the open government. Only a few countries, such as Canada, Mexico, Romania, and the UK have already included open contracting in their digital action plans. The open procurement and contracting approaches promises:

*   A single unified sourcing platform for government (Federal, State and Local Government) procurement.
*   Delivers transparency and accountability to government procurement processes.
*   Reduces procurement cycle times.
*   Reduces cost as a result of open, transparent and competitive bidding.
*   Increased procurement contract performance and effectiveness.

KEY BENEFITS OF SHARED TECHNOLOGY SERVICES IN THE NIGERIAN PUBLIC SERVICE

Shared technology services are innovative ways of organising administrative functions to optimise the delivery of cost-effective, flexible, reliable services to all stakeholders. Interest in shared services is rising globally, driven by the need to squeeze more value and costs out of organisations.

FINANCE	- General ledger - Accounts payable - Internet audit	- Accounts receivable - Purchasing - Insurance	- Tax compliance - Cash management - Foreign exchange
HUMAN RESOURCES	- Payroll processing - Compensation administration	- Benefits administration - Training & education	- Relocation services
INFORMATION SERVICES	- Standards - Technology /development	- Applications development - Applications maintenance	- Telecommunucations - Hardware & softwaare acquisition
LEGAL	- Litigation support and coordination	- Environment, health, and safety consulting / auditing	- Regulatory compliance
CORPORATE AFFAIRS	- Communication services	- Media relations	

Figure 19: Shared Services candidates [56]

Shared services are therefore the first class way of maintaining public services at a time when budgets are being frozen or cut, while simultaneously increasing motivation and job satisfaction for civil servants. The model is straightforward and based on a simple fact – the reason for the existence of the various government MDAs is to enable the efficient/effective delivery of frontline public services. It does include the need for the different back office or support services as it is currently the case. Instead, the administrative, back-office and support services can be shared as a common platform between government MDAs. The clear advantages of the paradigm shift are as follows:

1. Increased Efficiency and Reduced Costs

This is achieved through centralising dispersed operations and gaining economies of scale, and it also creates the opportunity to locate the operation in a low (or relatively low) cost location to reduce staff and infrastructure costs as well as improve labour mobility.

Increased Efficiency
- Economies of scale
- Technology leverage
- Standardization/coordination
- Reengineering opportunities
- Greater spans of control

Increased Effectiveness
- Specialization / skill leverage
- Free up management to focus on business issues
- Sharing information and resources across business

Lower Costs
- Increase profits
- Increase ROI

Achieve Strategies
- Increase revenues
- Increase market share

Figure 20: Shared Services Benefits [56]

2. Process Uniformity

Having central control of critical operations means one can more easily update standard procedures across the public service. This allows the government to adopt international best practices and in turn, can drive efficiency and improved service delivery across the federation.

3. Improved levels of Service Delivery

Often cited as a key outcome of a shared services approach, centralising critical operations allow the government to adopt best practices in delivering services internally and externally, respond to changing requirements, and undertake continuous improvement.

4. Increased Professionalization

Centralising specialist services often give a better opportunity to focus on the training and skills deployed in their delivery, which can often be overlooked when dispersed around the service.

5. Improved Career Prospects

Centralising teams focused on the main business processes often results in better career development opportunities for public servants, boosting motivation and driving up skill levels.

6. Better Technology

As well as improving processes and professionalisation, centralising services often mean that government can replace multiple dispersed technology systems with a single, modern one which would exploit the latest technological trends, local content development and contribute to improved service levels and efficiency.

DIGITAL TRANSFORMATION WITH CURRENT PUBLIC

SERVICE REFORMS

The Steering Committee on public service reforms based on the recommendation of the ministerial committee chaired by the Secretary to the Government of the Federation has led the development of a 10-year strategy for reforming Nigeria's public service. The strategy is founded on four pillars each led by a named organisation.

These were:

- Improving the Governance and Institutional Environment (led by the Office of the Secretary to the Government of the Federation);
- Improving the Socio-Economic Environment (led by the Ministry of Budget and National Planning);
- Improving Public Finance Management (led by the Ministry of Finance); and
- Improving Civil Service Administration (led by the Office of the Head of Civil Service of the Federation) [58]

The National Strategy for Public Service Reform of 2015 (NSPSR) acknowledges that some of the development challenges faced in Nigeria - *poverty, unemployment, security, and poor infrastructure* can be, to an extent, attributed to the weak performance of the public service. It acknowledges that the public service plays an essential role in the delivery of public goods and services and in creating an enabling environment for the economic and social development of the country. It, therefore, asserts that the dysfunction in public service administration undermines development and service delivery efforts at all levels and in all sectors of government as well as in the private sector [58].

A change in purpose changes a system profoundly, even if every element and interconnection remians the same[59]

The planned NSPSR's objective: "The starting point and core objective for all public service reform are to improve delivery of core services: all reform efforts must contribute to this goal. Reforms should be focused on achieving results regarding service delivery and not just on changing internal processes. Reforms must, in some way, make a tangible improvement in services delivered to citizens and other stakeholders, whether through improvements in quality, timeliness, access, equity and client focus."

Also, the mission statement of Nigeria's Federal Civil Service Commission reads, "*To ensure quality service delivery through efficient and effective implementation of government policies and programmes, working collaboratively and transparently with other stakeholders.*"

Digital transformation in public service, therefore, integrates well into those above and incorporates the following high-level cross-cutting objectives identified in the NSPSR strategy: "*To make a tangible improvement in services delivered to citizens and other stakeholders, through a focus on quality, timeliness, and equity and client orientation.*" [58].

The four pillars of public service digital transformation and NSPSR each have a more precisely defined development objective:

- To create a governance and institutional environment that enable public service institutions to deliver public goods and services by their mandates, and with integrity, transparency and accountability;
- To create a socio-economic environment that enables accelerated and sustained economic growth and poverty reduction, through

institutional pluralism and inclusive participation in the delivery of public goods and services;

- To achieve strategic, efficient and effective mobilisation, allocation and use of public resources, fiscal discipline, and transparency, integrity and accountability through timely reporting;
- To reinvigorate and transform the public service of the Federation as the primary institution for efficient and effective delivery of government policies and programmes.

Digital transformation will make tangible contributions to each of the goals listed in the national reform strategy document and more.

Case Study: Common procurement solution for London schools saves £300 million

The shared service for schools' IT in London's 33 boroughs and beyond, the London Grid for Learning (LGfL), has saved £300 million by aggregating procurement through shared services.

By using a joint system based on common standards, it has reduced duplication and increased efficiency for users, including processing 220,000 e-admissions for primary, secondary and early years so far this year.

Chief Executive John Jackson argues that the alternative, for similar organisations to independently develop their own solutions, is no longer viable. "Imagine trying to process those in 33 different ways. How much would that cost? How difficult would that be," he says in his presentation. "Why do things over and over again when you can do it once brilliantly, better and cheaper?

"Think of the paradigms that reshape the world – uber, Funding Circle, Bitcoin – all connect things together that weren't previously connected and enable sharing, are ubiquitous, are based on standards, are easy to use and they talk to each other," says Jackson. "That's what's transformational, that's the blueprint."[57]

"Everyone is a genius, but if you judge a fish on its ability to climb a tree, it will live its whole life believing that it is stupid"
— *Albert Einstein*

Chapter 8

·······◆◆◆·······

DIGITAL TRANSFORMATION SUCCESS FACTORS

Successful digital transformations from research uses a standard set of elements – see the digital transformation framework below. Each element is an enablement level to initiate and execute a digital transformation in any given situation, especially in government. These elements work together in an iterative way – always communicating and listening to re-envision and further re-enforces digital services provisioning. Senior public servants should drive digital transformation through an iterative three-step process as shown below [60]:

1. Envision the digital future for the public service.
2. Invest in digital initiatives and skills development.
3. Lead the change from the top and achieve stakeholders' buy-ins

"We've moved from digital products and infrastructure to digital distribution and Web strategy to now into more holistic transformations that clearly are based on mobile, social media, Digitalisation and the power of analytics, and we think it's really a new era requiring new strategies." – Saul Berman

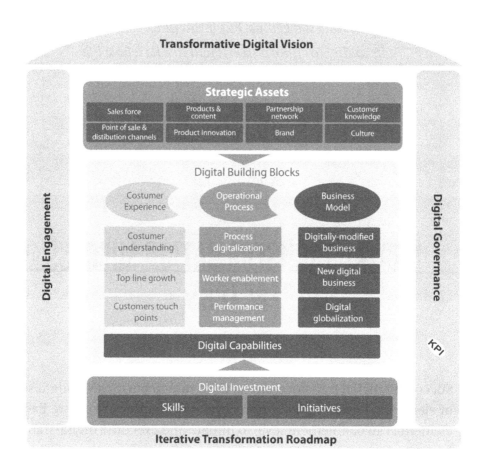

Figure 21: Digital Transformation Framework [60]

ENVISION THE DIGITAL FUTURE FOR THE PUBLIC SERVICE

Many digital transformation initiatives in government fail to capture all of the value available to them because their vision is not trans-formative. Nigeria's public services should be shaped around people's needs and be easily accessible, while also being delivered effectively and efficiently. Evolving Digital technologies offer great potential for Nigeria public service to provide wide-ranging advances on both of these fronts. Today, the majority of our people are accessing the internet through their mobile phones, laptops and tablets to receive information and to find services on the move, where they exist of course.

Crucially, new technologies not only allow greater scope for people to do things themselves, but also to contribute opinions, access information and interact with each other. There is significant poten-tial to transform how the public sector operates, with entirely new services and products becoming possible. Online delivery can make public services easier, quicker and more convenient for people to use, while also being less expensive for the government. Successful digital transformation does not occur only top down but also bottom up. A genuine value of transformation often comes from seeing value across silos and then helping everyone else see that value [60].

STEPS TO ENVISIONING

Step 1: Identify and diagnose existing strategic and functional assets: This step is about firstly getting to know what existing strategic and functional assets are in place with the mindset of envisioning, having a decent idea of what to be done. Also, it involves diagnosing them to take it up from there and select the best option possible.

Step 2: Leverage Existing Key Capabilities and Collaboration: Also, this is the step that closely follows the identification and diagnosis - being about leveraging the critical capabilities already in place to direct the purpose towards them.

Step 3: Define a step-wise transformation vision with clear delivery: Once the available capabilities and collaboration have been determined and leveraged, the next step to envisioning is to define a transformation vision with a perfect and precise delivery.

INVEST IN DIGITAL INITIATIVES AND SKILLS DEVELOPMENT

Transformation does not happen by chance or accidentally but with investment and dedication. Digital transformation is no different. To ensure we have a skilled and empowered workforce that delivers high quality digital public services, we have to develop a national or state approach to skills development for both the technology workforce and the wider public sector workforce. As with any investment, digital transformation requires understanding the need for investment, managing risk, and making the changes necessary to capitalise on the change. Also, there is often tremendous value to be gained from making the most of the investments through shared services.

The age of digital experimentation is over - with digital continuously showing healthy growth in the bleak landscape of slow economic recovery. The government then needs to move beyond experiments with digital and transform themselves and the public service into a new digital economy and business. Truly, digital transformation is uniquely challenging with it touching every function and business unit as well as demanding rapid development of new skills and investments which are very different from business as usual. To achieve success, the government needs to move beyond the vague statements of intent and focus on the actual digital in the public service system, structures, processes, and incentives. No blueprint for success, there has to be an investment into the digital initiatives and skills development. The process of investment, however, involves being unreasonably aspirational with the target, and acquiring capabilities (the skills required for digital transformation). The right talents should be hired even if they have to cost some hard cash, of course, because this is very vital to the success, at least in the very early stages of transformation. It is observed that significant lateral hiring is required to create a pool of talent deep enough to execute an ambitious digital agenda and also, sow the seeds for a new culture. Also, digital talent and initiative must be nurtured to ensure rapid improvements. Furthermore, rather than just accepting the historical norms, the government has the responsibility of challenging the status quo.

A lesson from the Scottish Government Digitalisation effort is the Digital Champions Skills Development Programme, a development programme for Chief Executives/Directors from across the public service. The purpose of this program is to inspire leaders about the transformational potential of digital tools and technology and to give them the confidence to take action to release their potentials. Once they have completed the formal programme, Champions are invited to join an alumni group to continue the networking and learning regime necessary to support further growth.

Another is the Scottish Top Talent programme, an innovative programme that is designed to nurture the skills and confidence of future public and third sector leaders to effect positive change in their organisations. Similar development programmes for youths interested in championing and supporting the digitalization vision is also appropriate in many climes. All these efforts must be properly aligned and supported by private sector as well [61]

The speed and breadth by which cities absorb and deploy technology, supported by agile policy frameworks, will determine their ability to compete in attracting talent. Possessing a superfast broadband, putting into place digital technologies in transportation, energy consumption, waste recycling and so on, help make a city more efficient and liveable, and therefore more attractive than others[3].

STEPS TO DIGITAL INVESTMENT

Step 1: Find and retain the right skills and talent: As always, it's hard to get the right expertise and talent for the digital transformation process, it has to be about finding the right hands for the job first. This process is very crucial because it helps to determine the future of the envisioned program.

Step 2: Invest in initiatives that advance the vision: After getting the right skills and talent, investment (monetary and otherwise) is the next thing to do. This means going as far as possible to breed the initiatives to advance the vision.

Step 3: Create sustained competition and forward thinking: The creation of sustained competition and forward thinking to keep the purpose alive is the last step to investing in digital transformation. This has the benefit of making sure it is not just a one-time thing, rather it is a process that is long term and has to be reviewed per time.

The purpose of the Digital Champions Skills Development Programme as mentioned is to inspire leaders about the transformational potential of digital tools and technology and to give them the confidence to take action to release that potential and act as guiding angels for sustainability.

A new, collaborative programme to train hundreds of Digital Champions to help thousands of people across the United Kingdom to get online was officially launched February 11, 2016. At a time when the United Kingdom is moving to 'digital by default' but over 12 million people lack basic digital skills[1], One Digital is a unique collaboration between six diverse public service organisations to create a sustainable and far-reaching digital inclusion solution. Age UK, the Scottish Council for Voluntary Organisations, Citizens Online and Digital Unite with its partners Affinity Sutton and AbilityNet have joined forces to develop the One Digital programme which will recruit, train and support 1,400 Digital Champions nationwide.

Funded with £2 million from the Big Lottery Fund, the programme will help around 9,500 people develop basic digital skills and reach a cross-sector of society including those with disability and accessibility needs, young adults looking for work, the over 65s, and third sector organisations and their beneficiaries. As the new One Digital Infographic demonstrates, Digital Champions are a useful tool in today's digital inclusion landscape where those who aren't online are now becoming harder to engage and need long-term personal support. Indeed 77% of those offline cite lack of interest as their primary barrier and 26% of computer beginners do not use their new digital skills without on-going help [62].

With 81% of learners rating their Digital Champion support as good or excellent, and 9 out of 10 Digital Champions saying they made a difference to getting people online, providing effective Digital Champion training and support can have a tremendous positive effect in engaging people with today's digital society.

Each Partner will deliver individual and bespoke projects, but all of the Digital Champions within the One Digital programme will access one single training and support network that provides structured courses, extensive resources and dedicated mentor support.

Underpinning the network will be a range of evaluation tools to capture essential impact data right across the partnership. The collaboration will also enable the partners to share learnings so that best practice models can be delivered for the benefit of all Digital Champions and end-learners.

Emma Weston, Programme Director for One Digital, said: *"Over the last two or three years, the digital inclusion space had changed significantly. It is no longer in question that digital skills are essential, but the digital skills issue can no longer be solved by third parties working in isolation. The One Digital partners have come together to pool our collective reach, expertise and skill to achieve a greater impact in the delivery of digital skills. Experience and evidence have shown us that empowering, supporting and inspiring people in our communities to be Digital Champions is a highly efficient and sustainable way to deliver digital skills. The One Digital programme will provide the vital rigour and infrastructure to the Digital Champion proposition so that meaningful digital inclusion can be achieved for everyone across society, now and for the long term."*

The role of a Digital Champion Volunteer would mainly involve:

- Engaging with residents to illustrate the benefits of basic computer skills and being online, and to signpost them to local support where appropriate.
- Encouraging residents to use the Self Service Portal and Self-Service mobile applications.
- Offering one-to-one support to individuals or groups.
- Attending training that would help you carry out this role.

WHAT SKILLS DOES A DIGITAL CHAMPION HAVE?

- A certain ability to use computers, search the Internet, e-mail, use social media, e.g. Facebook, Twitter, YouTube.

- Confident in a range of digital technologies, e.g. PCs, laptops, smartphones, tablets and enthusiastic about the benefits of digital technology and able to explain in a simple, clear way.
- Good listener with excellent communication skills.
- Willingness to accept any relevant or necessary training
- Awareness of people's interests and motivations for learning.
- Approachable and presentable, with an ability to empathise with diverse groups.
- Reliable and well-organised.
- Good sense of humour and willing to have fun.
- Able to travel to a variety of local venues, e.g. community centre library, etc.

PROGRAMME OBJECTIVES

1. Improve outcome transformation delivery, faster and more efficient uptake of digital tools and technologies within their spheres of influence.
2. Become early adopters of innovative digital concepts into their organisations and facilitating their effective application.
3. Adopt key user position to challenge current ways of working, and inspire new ways of working within their organisations and sectors.
4. Become change agents of how Digitalisation can improve outcomes for people, processes and approaches.
5. Create a critical mass of digital champions which builds a level of momentum around digital to achieve the Digital transformation objectives.

BENEFITS OF THE DIGITAL CHAMPIONS SKILLS DEVELOPMENT PROGRAMME

The benefits to the government are:

1. Nurture, develop and retain valuable leadership talent
2. Drive behavioural and sustained cultural change
3. Speed up delivery of change and service transformation projects

To ensure we have a skilled and empowered workforce that delivers high quality digital public services, we are developing a national approach to skills development for both the ICT Workforce and the wider public sector workforce. The benefits to the individual are:

1. Become a more confident, effective leader
2. Improve focus on leading people, not process
3. Be more successful in their personal ambitions

LEAD THE CHANGE FROM THE TOP AND ACHIEVE STAKEHOLDERS BUY-INS

The transformation of public service is accelerating rapidly with the availability of digital technologies, and the drive to reduce costs while improving services for citizens globally. The key to successful change is people and effective leadership. Top-level vision rarely translates to bottom-level action unless reinforced through top-down directives and governance. Excellent leadership skills and consistent engagement, backed with appropriate coordination, KPIs, and incentives, make the difficult process of digital transformation possible.

New and improved digital services require leadership and collaboration across many functional groups, including stakeholders from various government MDAs, programme management and technology functions.

Step 1: Engage Key Stakeholders

Public sector organisations thrive when they engage effectively with their stakeholders, but before they can do so, they need to know who those stakeholders are. In the broadest sense, a stakeholder is: *"A person such as an employee, customer or citizen who is involved with an organisation, society, etc. And therefore has responsibilities towards it and an interest in its success."* (Cambridge Dictionaries Online)

"Stakeholder engagement is the process by which an organisation involves people who may be affected by the decisions it makes or can influence the implementation of its decisions. They may support or oppose the decisions, be influential in the organisation or within the community in which it operates, holds relevant official positions or be affected in the long term."

New digital technology has transformed the way the public sector can engage with stakeholders. When you need to inform, consult or collaborate with individuals or organisations, you no longer have to rely on expensive or time-consuming techniques such as hard-copy surveys, face-to-face meetings, press adverts and interviews.

While these traditional methods are still valuable, taking stakeholder engagement online allows you to reach a larger audience, more cheaply and more often. While digital channels can open new opportunities for stakeholder engagement, their success is highly dependent on a supportive culture within the organisation. So if you plan to introduce digital stakeholder engagement to your organisation, or extend its use further, you will benefit from these tips for creating a supportive culture. They're based on over a decade's worth of experience helping the public sector use agile software to engage with others.

1. Put stakeholder needs and engagement outcomes first

There are many different tools and channels available to help you engage with the parties concerned, but be careful to adopt solutions that put stakeholder needs first — rather than the convenience of your teams. For example, if you need software that allows you to publish surveys online, your priorities might include:

- Cost – a low price per user
- Ease of administration – an intuitive interface for creating surveys
- Ease of use – to cut staff training costs
- Corporate templates – to create surveys under your brand.

While all of these priorities are laudable, there are a further two priorities that are more important:

- Ease of completion – stakeholders must be able to access and complete the published survey without special skills or training.
- Analysis of outcomes – the software must make it simple for you to analyse collected data. Otherwise, it will become an obstacle to understanding the results of your engagement.

Therefore when choosing digital channels or tools always put stake-holder needs and engagement outcomes first. If you don't, you and your colleagues will find yourselves frustrated by poor results when you engage with others.

2. Create a clear stakeholder engagement strategy

Don't incorporate digital tools and channels ad hoc into your general stakeholder engagement activities — create a clear stakeholder engagement strategy that incorporates both digital and traditional methods of engagement.

A clear strategy enables you to match the most appropriate digital channels to your work of informing, consulting and engaging with stakeholders. Once you've done this, you'll have a better idea of the software solutions you'll need, helping you to choose the right tools for your work. You'll also avoid wasting money on inadequate or overly complex software, which will, in turn, encourage staff buy-in and win their overall support for digital engagement methods.

If you need help mapping digital channels to stakeholder groups, take a look at the matrix in Transforming Stakeholder Engagement — Effective Digital Channels — it'll help you compile a list of channels you need for your work.

3. Choose software approved for public sector use

When you engage with stakeholders, you frequently gather confiden-tial data — and it's your duty to keep it safe. It could spell disaster if a cloud software provider didn't keep the data safe, damaging trust in your organisation, discouraging people and organisations from engaging with you, and damaging the culture of digital stakeholder engagement you are trying to build.

To prevent this from happening, choose software that — at a minimum — meets approved Government security standards. One way to make the due diligence work less onerous is by choosing software available on the UK Government's G-Cloud CloudStore that has already been pan-government accredited.

4. Shortlist and trial stakeholder engagement software

If you are selecting a cloud solution such as an online collaboration package, be sure to shortlist the software that best meets your needs (and those of your stakeholders) — and then try each one in turn. Many software packages — particularly those available via the G-Cloud Cloudstore — offer free trials, allowing you to put them through their paces at no financial risk.

When you try each solution, be sure to involve not only colleagues who will use the software, but management representatives as well. In this way you'll win support and foster understanding across the whole organisation, leading to a smoother and more positive adoption of the new systems.

5. Provide training and mentoring

When colleagues become accustomed to familiar ways of working, they can become resistant to change. Unless you support them in the new ways of engaging stakeholders, you will encounter reluctance in using them.

Choosing easy-to-use software will help you go a long way to addressing this problem, but also be sure to provide support and mentoring where it's needed. This will encourage adoption of the solution and can also generate new ideas and methods for engaging with stakeholders digitally — firmly embedding a culture of digital engagement among those responsible for reaching out to stakeholders.

Levels of stakeholder engagement

To engage all four groups of stakeholders, there is a three-stage hierarchy of engagement that allows you to inform, consult, and collaborate according to the type of stakeholder. Some stakeholders will require two or more levels of engagement, others only one. Informing stakeholders can inspire them to respond and emerge as valuable consultees, and consulting with them can see them become useful collaborators.

Informing and consulting can also keep high influence/low-interest stakeholders satisfied, and maintain the door open to collaboration if they should choose to get more deeply involved. Empowering stakeholders in this way becomes simpler using digital methods of engagement, as you will see later in this guide.

Most organisations would like their stakeholders to get more involved, which can only be done through improved communications and more opportunities to take part. Some stakeholders may only wish to be kept informed, while others may want to provide feedback and get more involved by volunteering their skills and resources. This pipeline effect ensures that you don't just rely on the famous 'few' and that all stakeholders have the opportunity to get involved in the ways that work best for them and you. The three main levels of engaging stakeholders require different approaches if they are to succeed, and each presents its challenges.

Step 2: Establish Governance

The dawn of the digital era is upon us, and as governments across the world begin to embrace digital technology, the way in which public services are delivered is changing dramatically. Recent research has found that progress varies not only between countries but also between domains within the public sector. Factors such as skills, leadership, culture and strategy all impact on digital maturity and can help, or prevent, governments accelerate the rate of their progress.

Deloitte's global research provides a digital maturity model, measuring the extent to which digital technologies have transformed an organisation's processes, talent engagement, and service models. The study identified that digitally mature organisations have key things in common.

Clear strategy aimed at fundamental transformation

An organisation's digital maturity is influenced, to a great degree, by its digital strategy. Among respondents from government agencies at the early stages of maturity, only 14% say that they have a clear and coherent digital strategy. In the case of more digitally mature organisations, the number grows sixfold, to 86%.

Digitally savvy leadership

The changes behind digital transformation challenge established models of leadership and governance. Before the ascent of digital technologies, new projects could be assessed through exhaustive analysis, investment decisions could be based on cost-benefit guidance, and the end destination of most plans was a fixed point. In the new digital era, leaders are required to make decisions more quickly in the face of a constant evolution in the art of the possible. In this challenging environment, just 38% of survey respondents believe their leadership has sufficient skills for digitally transforming public services.

For public bodies across the globe, the hierarchies and governance structures are often more pronounced than in the private sector. More than half of the respondents say a single person or group leads their organisation's digital agenda. Nearly 80% of these leaders are heads of various departments or agencies in governments, C-suite equivalents, or executives just below the C-suite level.

Greater user focus

Maturing organisations are nearly twice as likely as early-stage ones to be driven by customers/citizens' demand for digital transformation.

A laser focus on using digital technologies to improve the citizen experience helps maturing organisations improve service delivery. Respondents from evolving institutions almost unanimously report that digital technologies and capabilities enable their employees to work better with customers or citizens; in early-stage agencies, only a little over half the respondent says so. Additionally, 94% of maturing organisations have a digital strategy aimed at improving customer/citizen experience and engagement, compared to only 55% of early-stage ones.

Deeply embedded skills and culture

We asked survey respondents to rank the areas of digital transition that they deemed the most challenging to manage. Overall, workforce skills are the most challenging dimension of digital change. While culture comes second, responses are weighted toward culture being a particularly difficult area of change. So while 34% of respondents say that changing culture toward digital transformation was challenging, a high proportion of those characterise it as highly challenging. In other words, respondents recognise the level of change needed to ensure a digitally savvy workforce, but they understand that changing culture is a uniquely difficult task.

Procurement processes are muzzled by regulations and lack of flexibility

To deliver digital transformation, public bodies need to access a robust and innovative technology marketplace, but our survey suggests that procurement's capabilities fall well short of what's necessary to make that happen.

Seventy-six percent of respondents insists that procurement needs to change significantly or very significantly to accommodate digital transformation. When asked to rank the most significant obstacles to better procurement practices, they predominantly cite rules and regulations, lack of flexibility, and a lack of procurement skillsets.

Across the globe, the expectation of citizens and businesses has started to rise. We see the ease with which we can now bank, shop and travel and do not accept backwards public services – especially with welfare burdens increasing in the wake of economic slowdown.

At the same time, change is being forced on Government by falling budgets. The truth is inescapable: digital government can deliver the same outcomes for less money. Citizens demand digital services, and Government can't afford to maintain the status quo anyway.

The opportunity for digital transformation is to deliver a solution to both citizen expectations and the cost challenge in a unified way. Many people have thought that the technology to achieve this is hard, but survey has shown that the most difficult aspects of change are not the technologies at all, rather they are to change the culture, skills, governance regimes and commercial approaches. Leading public bodies have made significant headway and are now achieving results that are examples for others to follow.

The role of management in digital transformation

There is a host of new challenges facing the digital sphere – from the need for faster-integrated business cycles to increased calls for cross-silo capabilities, all the way through to the ever growing number of risks such as the potential for confidentiality and regulatory breaches. So how do we address these digital challenges?

The solution, according to a recent Capgemini research report, is digital governance. However, how exactly can public organisations improve their digital management capabilities?

What are the elements needed to make digital governance work?

There are some mechanisms that have been outlined to improve digital governance. However, here are the most common:

Shared digital units

One key factor for many governments looking to implement digital transformation is to create shared units that are used throughout the firm. According to the Capgemini report, shared digital units can be used to: develop digital services and develop new digital services.

With shared units, it's possible to reduce the costs typically associated with digital transformation because you are reducing the number of redundant initiatives needed in a local unit, and you don't need as many different areas of technology or as many people. Many companies choose to create a catalogue that outlines how to avoid redundant activities – and it may also be possible for them to carry out corporate wide investments that aren't feasible for individual units.

In addition, the same study reveals that missing skills are the biggest barrier towards making a digital transformation work. However, with shared units, it's possible to get the right people in place and then allow them to work closely with others and pass these skills on.

Governance committees

Another possibility is to establish an actual committee for digital transformation in the workplace. For example, a steering committee can aim to answer questions such as how to prioritise certain digital initiatives; how to allocate resources; and what rules need to be adopted to ensure that consumer experience is always consistent

Less common would be an innovation committee which could look to answer questions around emerging technology. Bringing people together around the table who have different viewpoints and different perspectives help to resolve issues and open up new ideas.

New Roles

Another possibility with digital governance is to establish specific leaders who will lead certain digital aspects of the business. For example, a chief digital officer may take responsibility for the web, social media, mobile and e-commerce overall – while there may be individual managers within each of these sections.

How do you make these elements work?

The main elements of digital governance must follow a number of key principles. Here is a brief rundown of the different paths you should follow.

1. Don't slow things down - Governance doesn't mean slowing things down with endless analysis. If anything, governance should mean the opposite – you should look to unblock anything that might be slowing down delivery and make decisions that help to increase the pace of delivery. Governing well actually means helping the team to avoid external pressures.
2. Make decisions at the right time - Focus your decisions on meeting the needs of the user and ensure that the team itself is empowered to make decisions so that a higher level is only called on when absolutely necessary. Short but regular meetings can be an ideal way of keeping on top of what's going on without slowing down the pace of delivery.
3. Find the right people - Remember that nothing will work without motivated, empowered and honest people. Make sure they are focused on key performance indicators and goals and that they are rewarded accordingly.

4. Be visible - While it's important to give your team room to manoeuvre and control their own actions for the bulk of their working time, make sure they know you are there. Regular meetings and visits are one way to achieve this – make sure you are on hand for demonstrations of progress and offer any feedback you deem necessary.

5. Focus on value - Make sure your team is focused on delivering value to the end user. There should be a clear vision in place and ways to measure success. Good governance means giving your team the opportunity to develop and explore any ideas that could potentially add value to the product, while also ensuring ideas don't go too far if they don't work.

6. Have trust - If you support the individuals in your team then have trust in them. Place the power in their hands as much as possible while still monitoring their progress. Focus primarily on outcomes and offer suggestions for improvement rather than trying to radically change things.

Digital governance is about taking control of projects to ensure they are focused and that any potential hitches can quickly be ironed out. With a motivated and talented team in place, governance should not be focused on pushing a team to follow strict criteria – it should be about empowering that team, bringing out their capabilities and ensuring they have the tools they need to flourish. With the right leadership, government business can be transformed for the better – and the same is true for its digital approach.

Step 3: Establish a new Digital Culture

Critically, in the coming years, millennial will represent a major source of talent who have the skills that are needed. However, it will be a daunting challenge to attract, train and retain them. A study by Deloitte notes millennial feels that businesses are out of touch with their personal values, for example, by focusing more on profit than social good. According to a World Economic Forum survey, there are three key things millennial are looking for from their employer:

career advancement (48%), company culture (38%) and training/development opportunities (32%).

To make the challenge harder, organisations now have to recruit and operate with greater transparency. Peer-to-peer reviews allow more insights into the inner workings of an organisation. Glassdoor, the fastest-growing jobs and recruiting website, alone, has reviews about salaries, interviews, CEO and workplace for more than 435,000 companies. Employees, particularly millennial, do not want to work for organisations that are badly rated especially government ministries and agencies.

A successful strategy for recruiting and retaining this talent stands or falls on the support it receives from leadership. The impetus to set, develop and foster this digital culture has to come from the top. Yet relatively few public companies make the radical changes to their boards and leadership that are needed. Russell Reynolds recently found that of the 300 largest companies in the United States, Europe and Asia, 72% do not have board members with digital literacy the case of Nigerian companies can only be imagined.

Building a digital workforce goes beyond just recruitment and talent development – it also requires incumbents to think about enhancing their workforce in other ways. Government enterprises can tap the online gig economy for skilled on-demand workers, speeding up the process of finding and onboarding talent.

Digital success isn't all about technology: The 2015 Digital Business Global Executive Study and Research Project by MIT Sloan Management Review and Deloitte identify strategy as the key driver in the digital arena. Companies that avoid risk-taking are unlikely to thrive and likely to lose talent, as employees across all age groups want to work for businesses committed to digital progress. The report is online and in PDF form, with a Digital Business Interactive Tool to explore the data set.

Digital strategy drives digital maturity. Only 15% of respondents from companies at the early stages of what we call digital maturity — an organisation where digital has transformed processes, talent engagement and business models — says that their organisations have a clear and coherent digital strategy. Among the digitally maturing, more than 80% do.

The power of a digital transformation strategy lies in its scope and objectives.

Less digitally mature organisations tend to focus on individual technologies and have strategies that are decidedly operational in focus. Digital strategies in the most mature organisations are developed with an eye on transforming the business.

Maturing digital organisations build skills to realise the strategy. Digitally maturing organisations are four times more likely to provide employees with needed skills than are organisations at lower ends of the spectrum. Consistent with our overall findings, the ability to conceptualise how digital technologies can impact the business is a skill lacking in many companies at the early stages of digital maturity.

Employees want to work for digital leaders. Across age groups from 22 to 60, the vast majority of respondents want to work for digitally enabled organisations. Employees will be on the lookout for the best digital opportunities, and businesses will have to continually up their digital game to retain and attract them.

Taking risks becomes a cultural norm. Digitally maturing organisations are more comfortable taking risks than their less digitally mature peers. To make their organisations less risk averse, business leaders have to embrace failure as a prerequisite for success. They must also address the likelihood that employees may be just as risk-averse as their managers and will need support to become bolder.

The digital agenda is led from the top. Maturing organisations are nearly twice as likely as less digitally mature entities to have a single person or group leading the effort. In addition, employees in digitally maturing organisations are highly confident in their leaders' digital fluency. Digital fluency, however, doesn't demand mastery of the technologies. Instead, it requires the ability to articulate the value of digital technologies to the organisation's future.

Digital Transformation Isn't Really About Technology

AdvertisementOne wouldn't expect that changing the size of tables in an employee cafeteria could be emblematic of the digital transformation of business. But consider this example: The tables in question were in the offices of a large, online travel company working with Humanyze, a people analytics company headquartered in Boston that is a spinoff of the MIT Media Lab. Humanyze integrates wearable's, sensors, digital data and analytics to identify who talks to whom, where they spend time and how they talk to each other. The analysis identifies patterns of collaboration that correlate with high employee productivity.

Humanyze analysed the travel company's workforce and discovered that people eating lunch together shared important insights that made them more productive. In addition, the analysis showed that productivity went up based on the number of people at the same table. At the company being analysed, Humanyze found that employees typically launched with either four or 12 people. A quick inspection of the cafeteria solved the puzzle — all the tables were for either four or 12 people. The integration of digital technologies pointed the way to increasing table sizes, which had a direct and measurable impact on employees' ability to produce.

To understand the challenges and opportunities associated with the use of social and digital business, *MIT Sloan Management Review*, in collaboration with Deloitte, conducted its fourth annual survey of more than 4,800 business executives, managers and analysts from organisations around the world. The survey, conducted in the fall of 2014, captured insights from individuals in 129 countries and 27 industries and involved organisations of various sizes. The sample was drawn from a number of sources, including MIT alumni, *MIT Sloan Management Review* subscribers, Deloitte Debriefs webcast subscribers and other interested parties. In addition to the survey results, they interviewed business executives from a number of industries, as well as technology vendors, to understand the practical issues facing organisations today. Their insights contributed to a richer understanding of the data. Surveys in the three previous years were conducted with a focus on social business. This year's study has expanded to include digital business.

The tale of the tables is a powerful example of a key finding in this year's *MIT Sloan Management Review* and Deloitte digital business study: The strength of digital technologies — social, mobile, analytics and cloud — doesn't lie in the technologies individually. Instead, it stems from how companies integrate them to transform their businesses and how they manage work.

Another key finding: What separates digital leaders from the rest is a clear digital strategy combined with a culture and leadership poised to drive the transformation. The history of technological advance in business is littered with examples of companies focusing on technologies without investing in organisational capabilities that ensure their impact. In many companies, the failed implementation of enterprise resource planning and previous generations of knowledge management systems are classic examples of expectations falling short because organisations didn't change mindsets and processes or build cultures that fostered change. The report last year on social business found similar shortcomings standing in the way of technology reaching its potential.

The findings this year are based on an assessment of digital business maturity and how maturing organisations differ from others.

To assess maturity, they asked respondents to "imagine an ideal organisation transformed by digital technologies and capabilities that improve processes, engage talent across the organisation and drive new value-generating business models." They then asked them to rate their company against that idea on a scale of 1 to 10. Three groups emerged: "early" (26%), "developing" (45%) and "maturing" (29%).

To assess companies' digital maturity, we asked respondents to rate their company against an ideal organisation — one transformed by digital technologies and capabilities — on a scale of 1 to 10. Three groups emerged: "early" (1–3), "developing" (4–6) and "maturing" (7–10).

Although we found some differences in technology use between different levels of maturity, we found that as organisations mature, they develop the four technologies (social, mobile, analytics and cloud) in nearly equal measure. The greatest differences between levels of maturity lie in the business aspects of the organisation. Digitally maturing companies, for example, are more than five times more likely to have a clear digital strategy than are companies in early stages. Digitally maturing organisations are also much more likely to have collaborative cultures that encourage risk taking.

Several obstacles stand in the way of digital maturity; lack of strategy and competing priorities lead the list of speed bumps. Lack of a digital strategy is the biggest barrier to digital maturity for organisations in the early stages, according to more than 50% of respondents from early-stage organisations. As companies move up the maturity curve, competing priorities and concerns over digital security become the primary obstacles.

While a lack of strategy hinders early and developing companies, security issues become a greater concern for maturing digital companies.

Across the board, respondents agree that the digital age is upon us: Fully 76% of respondents say that digital technologies are important to their organisations today, and 92% say they will be important three years from now. In this year's report, which is based on a survey of more than 4,800 executives and managers as well as interviews with business and thought leaders, we look specifically at the emerging contours of digital business and how companies are moving forward with their digital transformations.

Creating a Strategy That Transforms

When developing a more advanced digital strategy, the best approach may be to turn the traditional strategy development process on its head. Benn Konsynski, the George S. Craft Distinguished University Professor of Information Systems & Operations Management at Emory University's Goizueta Business School, proposes that rather than analysing current capabilities and then plotting an organisation's next steps, organisations should work backwards from a future vision.

"The future is best seen with a running start," Konsynski comments. "*Ten years ago, we would not have predicted some of the revolutions in social or analytics by looking at these technologies as they existed at the time. I would rather start by rethinking business and commerce and then work backwards. New capabilities make new solutions possible, and needed solutions stimulate demand for new capabilities.*"

As an example, Konsynski points to the spice and flavour manufacturer McCormick & Company. Given the importance of personalization and digital technology's ability to provide it, McCormick developed FlavorPrint, an algorithm representing the company's flavours as a vector of 50 data points. Currently, McCormick uses FlavorPrint to recommend recipes to its consumers. But the vision is much bolder. McCormick thinks of FlavorPrint as the Pandora of flavourings, which has prompted the organisation to see itself as a food experience company rather than a purveyor of spices.

Eventually, all McCormick flavours will be digitised, and the company will be able to tailor them to regional, cultural and even individual personal tastes. Although all the needed technologies are not yet available, they likely will be in the coming years, and the strategy to take advantage of them is already in place. The FlavorPrint product has shown such promise that McCormick recently spun it off into its own technology company, Vivanda, with former McCormick CIO Jerry Wolfe as its founder and CEO. As it confronts changing consumer tastes, McDonald's is digitally revamping its restaurant experience and how the company works. The global restaurant chain was one of the first companies to adopt the Apple Pay mobile payments solution. Last year, it installed kiosks in select locations that allow customers to order customised hamburgers. And it's seeking partnerships with start-ups, such as a company that embeds sensors into the paper.

McDonald's is also integrating digital technologies to spur the organisation to work in new ways. It's ambitious campaign during the 2015 Super Bowl football championship is an excellent example: McDonald's planned to give away an item related to every commercial that aired during the game.

To respond to commercials almost instantaneously, McDonald's had to integrate multiple digital technologies and reconfigure its internal communication and operational processes. The integration came together in a digital newsroom with a cross-functional team that included members from the company's marketing and legal divisions, representatives from the company's various advertising agencies and employees from the company's enterprise social technology provider.

Meeting the goal required real-time reactions and monitoring and analysis of social media trends. It also demanded on-the-spot decision making to come up with the best decisions about which products to give away. The effort was successful and drew 1.2 million re-tweets, including some from high-profile celebrities such as Taylor Swift.

The event was part of an on-going effort at the fast food chain to transform themselves into an organisation that integrates technologies to become more agile, experimental and collaborative. As Lainey Garcia, manager of brand public relations and engagement at McDonald's put it: "The biggest takeaway was the power of integration. You can accomplish amazing things when you have all those pieces working together collectively in a holistic way."

Case Study: Everledger - uses blockchain to prove diamond authenticity

Everledger is solving a massive insurance problem – those associated with diamond theft using blockchain technologies. Every year millions of dollars of fake and fraudulent diamond theft transactions take place globally with studies showing that 65% of fraudulent diamond claims go undetected, leading to insurance companies bleeding money left, right and centre.

The issue is that there are few tools available to insurers to detect whether or not diamonds have or haven't been stolen. Proof of ownership of jewellery, just like precious metals and other high-value commodities, is locked in paper, and paper can be easily corrupted or misplaced.

Everledger offers a permanent ledger for diamond certification and related transaction history on the blockchain, and acts as verification for insurance companies, owners and law enforcement. In other words, it's a tamper-proof digital ledger for the world's diamond market.

Blockchain technology as proof of existence, proof of ownership, proof of transaction, proof of exchange, proof of value - any form of proof, is where this technology is showing its greatest usefulness. Replacing paper proof with digital proof is at the heart of the blockchain revolution, and it's where most blockchain start-up companies are focusing, and Everledger is leading the pack, as it has addressed a very specific issue with a smart solution [54].

"Commitment is what transforms
a promise into reality."
— *Abraham Lincoln*

Chapter 9

........................•◆•........................

DIGITAL TRANSFORMATION AND NIGERIA

Digital technology has fundamentally changed the way people live, work, interact and learn. Commercial entities have been quick to adapt. In Nigeria, this evolution has been witnessed in the private sector, for instance, the retail, telecommunication and banking industries have led the way in fulfilling their customers' demands through always-on digital channels. Going digital! Nigerian businesses are now embracing digital transformation which will give absolute insight into digital technology adoption in the Nigerian marketplace. However, there are certain factors that affect technology adoption in the nation as a whole - with institutions, infrastructure, vulnerability, awareness, labour, and innovativeness by local businesses all having a part to play in embracing and enabling the next big digital innovation.

In 2001, an e-government project was initiated when the National Policy on Information Technology (IT) "USE IT" was approved by the then President Olusegun Obasanjo. This policy document set out a strategy for digital government implementation in Nigeria. Tasked with the function of coordinating the project was the National Information Technology Development Agency (NITDA) working together with National e-Government Strategies Limited (NeGSt). The policy was aimed at integrating agriculture, health, education and other sectors aided by the Federal Government of Nigeria's data and research satellite launched in 2003[63].

The NeGSt was a tripartite joint venture, that is, a platform involving three distinct parties, namely, the government (represented by NITDA), private and financial investors, and technology partners, for which their shares of the ownership in the joint venture is 5%, 15% and 80% respectively[64]. The essence of this strategic alliance was to create a unified national framework for the adoption and implementation of digital technologies in and between government agencies and their clients. As stated on its website (www.negst.com.ng), the purpose of NeGSt's creation was "to facilitate, drive and implement the Nigerian digital government programme under a public-private partnership model [64].

Following the creation of NeGSt in 2007, one of its first initiatives was to establish an online database for teachers in Nigeria using digital registration [65]. In addition, NeGSt instituted and projected other digital governance projects in the country, for instance:

- Digital readiness was a programme designed to prepare the Nigerian civil service for e- governance.
- Digital parliament was an initiative to bridge the gap between the citizens and lawmakers via the Internet and mobile telecommunication channels.
- Digital passport for the preparation and issuance of international passports.
- Digital NYSC involved the administration and management of the National Youth Service Corps (NYSC) through the use of digital technologies.
- Digital Local Government councils were created with the purpose of making Local Government Areas (LGA) live up to their responsibilities as the closest representation of government to the people via the application of technologies [66].

At the time the *"USE IT"* policy began, Nigeria's capacity to implement digital governance was considered lower than countries such as Egypt, South Africa and Kenya. Nonetheless, the country experienced an impressive turnaround not long after the deregulation of the telecommunications sector which began in 2001 and soon resulted in the skyrocketing of mobile telephony and a significant amount of direct capital investment by foreign firms in the sector.

The country quickly overtook every other country on the continent principally due to the size of its population and the wealth circulating in its local economy which enabled a sizeable number of citizens, enterprises – and the government itself – to buy ICT equipment that included personal computers, smartphones and other Internet-related devices, and paved the way for Internet programmes that were meant to integrate governmental ministries and departments.

Despite this progress, Nigeria currently has an e-government development index of 0.2929 and stands at 141 out of 193 countries surveyed in 2014. This rating puts it slightly above the Western Africa regional average of 0.2660 but distinctly behind the regional champion Ghana, which has an index of 0.3735 and well behind the global average of 0.4712 [67].

There are two reasons to which this lack of progress can be attributed. First, the importance of digital technologies and their disruptive effect on government is not yet well understood nor applied in the public services. The potential of digitising public services for both the populace and the civil servant has not been fully explored by the government. Second – as a consequence of the first – those would-be digital leaders often do not have the mandate to drive changes. IT specialists are placed in faraway departments and are compartmentalised, often unable to manage the ICT function across government departments. Thus the potential for leadership in the digital process has been lost and wasted.

DIGITAL TRANSFORMATION OF THE NIGERIAN PUBLIC SERVICE CHALLENGES

The process of digitalization requires the Nigerian government to take digital transformation beyond the provision of online services through digital government portals, into the broader business of government itself. Digital transformation requires changes to both processes and technology systems which are more challenging to implement in the public sector than in the private sector.

The public sector must cope with additional management issues, including duplicity of agencies, a range of organisational mandates and constituencies, longer appropriation timelines, and the challenge of maintaining strategic continuity even as political administrations change. In addition, the size and demographic of the clientele – (the entire population of the country is the client base) a number much larger than that which most businesses serve, must be taken into account.

Similarly, when systems and or data are owned by different departments, agencies and functions, on a range of platforms, with differing taxonomies and access requirements, it can be difficult to invest at a large scale and generate sufficient economies not to mention expected master data management challenges. The complexity of large-scale digital transformation projects requires specialised skills and expertise that come at a high cost and are often in short supply. Silos, fragmentation, and the absence of a central owner for nationwide technology infrastructure and common components can make it hard to create a seamless experience for the end-user [68].

Contrary to claims by Galaxy Backbone Limited, an Information and Communication Technology services provider wholly owned by the Federal Government of Nigeria, no clear, verifiable data are available on how effective it has been on the ground in terms of pursuing and harmonising technology acquisition, operation and use in the public sector of Nigeria. Since its inception in 2007 like the public sector it was designed to serve, its mandates have continued to beg for more attention.

Although the objectives of this work are far ahead and above those of Galaxy Backbone Limited's mandate, the subset that agreed with the proposition of this book are as enumerated below. Galaxy Backbone was expected to support the objectives of the Federal Ministry of Communication Technology with the following:

- Build and operate a single nation-wide IP broadband network to provide network services to all Federal Government MDAs and institutions.
- Be the provider of shared ICT infrastructure, applications and services to all Federal Government MDAs and institutions, e.g. manage Government Data Centres and databases, Directory Services, National Information Repositories, IP-telephony and other solutions.
- Set standards and guidelines for government MDAs in the acquisition and acceptable usage of ICT infrastructure, applications and services across Federal Government MDAs and institutions.
- Be the provider of technical support to the Ministry of Communication Technology for an end to end Quality Assurance of ICT projects and capacity building for ICT professionals in Government.

Other observed challenges to digital transformation in the Nigeria public service are as follows:

Poor Leadership

Any society where public figures exemplify the sterling qualities of good leadership, invariably rub off on the society at large, and this encourages fellow citizens with the same patriotic fervour. However, a land devoid of visionary leaders and a nation without integrity can hardly experience stability, peace and development. Nigeria has been in recent times dominated by a worsening security situation, manifested in sporadic violent skirmishes, armed robbery, arson, politically motivated assassinations, massive youth unemployment and acrimonious poverty [67]. In contradiction to this national scenario is a political and civil leadership class that wallows in financial and economic constipation and social self-adulation.

The benefits of digital transformation in Nigeria and what it has to offer has been discussed front and back extensively. Technology scholars have talked about its implementation benefits in ensuring accountability, awareness and transparency in the management of governmental business. Also, an efficient, speedy and transparent process of information dissemination to the public is achieved and the performance of the administrative personnel and good governance, in turn, are encouraged. However, there are certain challenges that this faces. The over-bloated public service whose members will see and analyse this practice as an attempt by the government to relieve some members of their jobs, therefore, a number of the public servants are likely to frustrate its application. Also, Nigerians are hardly ever known to quit public offices on grounds of failure, incompetence, neglect, scandal or moral integrity. Even in the face of obvious neglect and mismanagement of public resources, it is not in our character to quit the stage, and when forced to do so following scandals, our leaders are let off the hook to flaunt their ill-gotten gains and to warm their way back to political relevance at various levels, using the same ill-gotten wealth to buy the peoples' support.

Infrastructural Deficiency

Infrastructure is the basic physical and organisational structure needed for the operation of a society or nation state like industries, buildings, roads, bridges, health services, governance, etc. It is the products, services and facilities necessary for an economy to function properly. The infrastructural development statistics of Nigeria just like other third world countries is nothing to write home about. Most physical infrastructure in the country are now decayed and need repair, rehabilitation or replacement. Governance is the process of using public services to plan, organize, control and supervise the people who are resident in a country (for example to have a conducive environment for living and a sense of belonging). Governments have the prerogative to put in place all measures that it deems fit to make an environment beneficial for living for everybody living there, especially for its citizens.

Infrastructural developments in third world countries like Nigeria are sometimes more challenging because of the accessibility of people to government. Another challenge is the process of identifying the right project, carrying out feasibility and viability studies and embarking on the physical development of viable projects. These challenges are numerous and include finance, technology for development, maintenance and design. These challenges also include quality requirements of projects to meet international standards and to be sustainably developed. Projects must also meet the carbon emission standard set by international organisations like the International Standards Organisation (ISO), United Nations Environment Programme (UNEP), etc. In other words, there is a tendency of lack of trained and qualified personnel to handle and operate its infrastructures in the Nigerian public service. Moreover, the high cost associated with the procurement and training of public servants with the ICT skills may cause some reluctant feeling on the part of the government in its implementation in the public service.

Dearth of Manpower

Historically, education experts have established a nexus between the perennial lip service paid by successive governments, and the glaring dearth of requisite skilled technical manpower to address the country's developmental challenges. The last Association of Telecoms Companies of Nigeria (ATCON) forum held in Lagos in 2015 for instance, the President of Association of Licensed Telecoms Operators of Nigeria (ALTON) – Engr. Gbenga Adebayo warned of the current manpower challenges observed in the industry which, if not addressed, the Nigeria state may be running a telecommunication network that will be professionally powered by foreigners in the next one decade or more [69].

The topic of manpower and know-how in digital transformation cannot be overlooked as well. ICT education is a very important issue and as highlighted by the Director of NCC in an interview, the fact that majority of staffs in the agency do not have the required level of knowledge in ICT to carry out the task of Digitalisation causing the very few with ICT knowledge to be overburdened.

The above warning is true generally in other technology-related industries as the inability to produce the all-important skilled manpower continues sometimes attributable to the poor perception of technical education, out-dated school curricula and the absence of a focused policy on digital technology development in the country.

Lack of Supporting Legal Framework

In Nigeria, there is a huge distrust between public services and the people. This has created a lot of tensions; so much so that when the Government launches a potentially great initiative or program, the lack of trust acts as a saboteur. The distrust being felt in Nigeria today is a natural consequence of state capture by the few and for the few. The way to start rebuilding trust is to allow people the means to verify the performance of public services. This is only possible where adequate mechanisms and an enabling legal framework(s) are implemented to ensure that people can participate effectively in holding the government to account.

This will engender a new era of robust ideas birthed by open scrutiny from different stakeholders who would be able to see things from different perspectives. With consistent engagement, the people begin to trust that the government is indeed representing their interests. Robust engagement also provides the opportunity for the governed to manage their expectations especially in terms of deliverables, setting objectives, etc. this builds trust which becomes a fertile ground for idea implementation. The public services also benefit from engaging with the people they serve because mutual understanding is further fostered and priorities are better directed at meeting the most urgent needs that are in the public interest. As a result, the public services and government earn the people's respect and support.

Scepticism and Ownership

Scepticism about emerging technologies is fuelled by the potential or actual misuse or abuse of what should be confidential data. In addition, misuse of infrastructure by public service staff and the potential for damage to technology infrastructure may add to the costs of maintenance and replacement. Without a sense of ownership, the public will not be interested in protecting and contributing to the maintenance of public technology infrastructure as is the case today.

Public Service Culture and Capacity

Although there are some examples of the civil service responding quickly and efficiently to new challenges, and innovating with new approaches to old problems. But this old culture can be cautious and slow-moving, focused on processes, not outcomes, bureaucratic, hierarchical, and resistant to change. This culture can make it difficult for a government to adapt to new demands from their populace.

Furthermore, the Nigerian public service though diverse and encompassing a myriad of professions, educational levels and experiences, they may not be in possession of the right skills required for the Digitalisation process at this very moment.

The Nigerian End-User

Despite enormous gains in Digitalisation in the private sector coupled with increased technical education and skills training, and the pursuit of the Millennium Development Goals (MDG) in education, a large number of Nigerians will remain unable to access digital services due to inadequate literacy levels – especially among the elderly, the poor and marginalised. In addition, Internet access has yet to reach many rural areas, urban and low-income areas, and geographically remote places. Furthermore, conflict and unrest, for instance in areas under siege by Boko Haram, may make it difficult to invest in and create the necessary enabling infrastructure for digital transformation.

DEFUSING CHALLENGES TO PUBLIC SERVICE DIGITAL TRANSFORMATION

International best practice [70] recommends the following actions should be pursued in order to define challenges to digital transformation:

i. Establish central coordination and commitment at all levels of government.
ii. Invest in human resource development.
iii. Avoid large-scale hiring and firing, rather a focus on identifying existing skills and capacity for skills upgrade are preferable.
iv. Make public service enterprise data and analytics available to improve decision-making in the government.
v. Protect critical infrastructure and confidential data.
vi. Encourage ownership of the process through stakeholder involvement and building citizen capacity to access the services.
vii. The digitalisation process must be integrated with wider public sector reforms.

THE NIGERIAN PUBLIC SERVICE AND THE CASE FOR DIGITAL TRANSFORMATION

The Nigerian public service dates back to March 1862 when the British government declared its interest in the Port and Island of Lagos. A new government was constituted, and provisions were made for key posts such as Governor and Chief Magistrate. By 1906, the British Government had extended its authority over most of Nigeria and began to establish its instruments of law and order such as the departments of judiciary, police, prisons, public works, and customs [17]. The structure of the service as it is known today was put in place by Sir Hugh Clifford who succeeded Lord Frederick Lugard and was appointed Governor of Nigeria. He established a central Secretariat in Lagos in 1921.

In 1939, similar secretariats were established for the three provinces which were administered from Ibadan, Enugu and Kaduna. The 1940s and 1950s saw the emergence of the nationalist Nigerian administrators. This period also marked the beginning of repeated pressure for reforms in the Nigerian political and civil service structure.

The military governments in 1966 to 1979 and 1984 to 1999 'dealt a blow' to the civil service. Nigeria experienced over 25 years of military rule with only short periods of civilian administration. These military regimes disbanded the legislature and did away with the system of elected citizen representation. It showed a lack of skill and expertise in managing Nigeria's economy and development.

The Public Service of Nigeria

The structure of the ministries reflects the mandates and responsibilities of the ministries. Each Ministry has a Minister as its political head being supported by Minister of State in some cases and a Permanent Secretary as the administrative head. The number of Departments (headed by Directors) ranges from four to eight depending on the functional composition and complexity of the Ministry. However Finance and Accounts (F&A), Human Resources Management (HRM) and Administration are the three groups of function which must be present in the structure of all Ministries.

The recruitment of Federal Civil servants is vested in the Federal Civil Service commission. A Chairman heads it. Its membership has recently been enlarged to fifteen.

Figure 22: Leadership in Nigeria Public Service

The military governments made and implemented policies, as well as enacted and enforced laws in the absence of elected representatives. They co-opted former politicians, academics and top civil servants to ministerial posts, giving them enormous powers to initiate and execute economic, social and political policies and programmes. These led to the breakdown of accountability for public spending and a lack of probity in the service. During this time, ostentatious lifestyle and brazen corruption became pervasive among civil servants and nearly everyone who had access to political power, which was especially through nepotism. It has been reported that between 1988 and 1994, approximately $12.5 billion in government revenues was syphoned out of the country using what has been termed "special accounts" or failed contract schemes [71]

The military regimes brought a sense of insecurity to the civil service. These were exacerbated by the practice of purging senior public service officers, a practice which was repeated by successive military administrations. Civil servants worked in a state of fear and uncertainty and an administrative culture of dehumanisation and frustration. They lacked the tools and equipment to fulfil their obligations, which, when coupled with the runaway inflation that the country experienced, kept an average civil servant in misery, while another consequence was a sense of apathy to professionalism. Persistent low morale, embezzlement and corruption ultimately resulted in ineffectiveness and low productivity in the service. Higher tier civil servants assumed enormous political power without an accompanying sense of duty or accountability to the people.

Case Study: UK Digitising Land and Property Searches

Until recently, the main way for people to find out whether land or property is registered was to go through a lengthy process called Search of the Index Map, or SIM. This entails filling in an application form, paying a small fee and then waiting a number of days to receive information about the desired property or piece of land.

In an attempt to drag this process into the 21st Century, a team of officials at the Land Registry put their heads together and came up with the "MapSearch" tool. This new digital service, which went live in March 2014, allows businesses to quickly establish whether land or property in England and Wales is registered, view its location and see title numbers and details of tenure. As a result of MapSearch, the number of SIM requests has been significantly reduced and the Land Registry has digitised a huge number of its paper transactions. In recognition of their work, the Land Registry team scooped the Digital Award at the 2015 Civil Service Awards. But getting from a blank slate to the completed MapSearch project was no mean feat [39]

"In the new world, it is not the big fish which eats the small fish; it's the fast fish which eats the slow fish"
— *Klaus Schwab*

Chapter 10

DIGITAL TRANSFORMATION AND THE NIGERIAN APPROACH

Around the world, governments are encouraging digital transformation through digital service enablement. Government digital service is becoming the norm and central to technology adoption in the public service globally. It also true that rich governments are exploring the possibility of using blockchain technology, an innovative idea that underpins the bitcoin crypto-currency to increase efficiency in the delivery of more efficient and trust-based services. A blockchain works as a decentralised ledger that is verified and shared by a network of computers, and can be used to record data as well as to secure and validate transactions. Banks and other financial institutions are increasingly investing in blockchain technology, reckoning it could cut their costs and make their operations faster and more transparent.

The concepts and structures developed for blockchains are extremely portable and extensible to other areas of economic and social activity. As such, they have a profound potential for application within government services — indeed the eventual impact may be a significant reduction in the cost of governance in Nigeria. The blockchain technology as the basis of government digital transformation of public service thoroughly addresses issues of privacy, security, identity and trust — distributed ledgers creates genuine opportunities for the government at all levels to reap immense benefits:

- Reduced cost of public services, including reducing fraud and error in payment transactions.
- Greater transparency of transactions between government agencies and citizens.
- Greater financial inclusion of people currently outside the former banking system.
- Reduced costs of protecting citizens' data while creating the possibility of sharing data between different entities, allowing for the creation of digital marketplaces.
- Protection of critical infrastructure such as schools, government buildings bridges, pipelines, etc.
- Reduced market friction, making it easier for small and medium-sized enterprises (SMEs) to interact with local, state and national governments.
- Promotion of innovation and economic growth possibilities for small and medium-sized enterprises (SMEs)

Blockchain distributed ledger systems have the potential to be radically disruptive. Their processing capability is real time, near tamper-proof and increasingly low-cost. They can be applied to a wide range of industries and services, such as public services, financial services, real estate, healthcare, and identity management. They can underpin other software and hardware-based innovations such as smart contracts and the Internet of Everything. Furthermore, their underlying philosophies of distributed consensus, open source, transparency and community could be highly disruptive to many of these sectors. Like any radical innovation, as well as providing opportunities, a distributed ledger creates threats to those who fail or are unable to respond. In particular, through their distributed consensual nature they may be perceived as threatening the role of trusted intermediaries in positions of control within traditionally hierarchical organisations such as government departments and agencies. With its wide range of stakeholders, services and roles, the government has a multitude of different operations. Some distribute value rather than create it, and others create and maintain effective regulatory regimes. Many of these activities will be enhanced by innovations afforded by distributed ledgers, and others will be challenged significantly. Ultimately, the best way to develop this technology is to adopt and use it in practice.

Blockchain undoubtedly holds value for governments in developing countries, offering new ways of operating that reduce fraud, error and the costs of delivering services to underserved citizens. At the same time, these technologies offer new forms of innovation and the ability to reduce transaction costs for small and medium enterprises in Nigeria. This chapter highlights only some of the possible use cases more are showcased in later chapters. As distributed ledgers are adopted more widely, it is likely that a new form of operating government services will emerge.

VISION, GOALS, THEME AND GUIDING PRINCIPLE

This book identifies specific changes needed across the Nigerian public service to enable service delivery and integration digitally. It sets out key action points, which are not exhaustive, and should be regularly updated and reviewed on a continuing basis during and after implementation or transformative growth phase.

VISION

A fully functional, and agile public service administration, resolutely oriented towards the Nigerian citizen, with adequately resourced core services, providing a nurturing and rewarding working environment for public servants and high-quality services to citizens.

The objective of digital transformation in the public sector should include:

To implement digital transformation in the public sector to empower institutions and improve public service delivery.

GOALS

The goals of the digital transformation process should include:

- To improve the coverage and quality of government service delivery.
- To promote equal access to government services through the use of modern digital technologies.
- To increase accountability and reduce corruption.
- To reduce transaction costs for citizens and the government.
- To enhance work-life balance of public servants.
- To improve the standard of living of the average Nigerian.

THEME

The programme's theme can be crafted as "*Towards a digital Nigerian Public Service.*"

GUIDING PRINCIPLE

The following key principles should provide guidance to Digital Transformation implementation leads:

- **Simplification**: Analyse opportunities to simplify and re-engineer core government processes under a common architecture and use common components wherever possible to enable transparency, visibility and auditability.
- **Reusability**: Leverage existing technology investments where possible.
- **Automation**: Automate repetitive and complex activities with due regards to workflows, governance, risk and compliance requirements.
- **Integration**: Integrate core businesses within the existing system, thereby guaranteeing a single source of business truth.
- **Information**: Provide a foundation for business intelligence using an appropriate big data solution and centralise all reporting requirements.

Digitalisation with representation is key - as it cuts across all spheres (public and private). It must include more than just the government top officials and the elites. Risk, compliance, partners, legal, and IT need to be involved in the governance hierarchy for the successful implementation. It is important to understand that leadership (not technology) delivers successful digital transformation. In simple terms, doing digital transformation requires the merging of technology and marketing, ensuring customer-focused digitalisation, using the process to ensure structure, management giving pointers and encouragement, responding to change when confronted and striving for likes.

DIGITAL TRANSFORMATION IMPLEMENTATION APPROACH

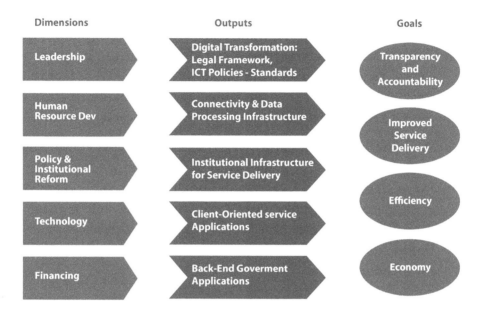

Figure 23: Digital Transformation in the Nigeria Public Service

The digital transformation implementation approach should complement on-going public sector reforms through digital transformation of the civil service and its public service delivery functions in a proposed 48-month timeframe:

- Adopting and implementing the public service digital transformation roadmap as discussed in this book.

- Undertaking a digital capability assessment, digital services mapping and planning in the public service and establishing the extent and nature of digital service provision.

- Facilitating the establishment of functioning digital services led implementation agency and providing the initial capacity building required.

- Carrying out a coherent roll-out, at national (federal) and sub-national levels (state and local government councils), of citizen-centred digital services which simultaneously underpin the legitimacy of government action.

- Creating a time-bound lead agency to manage the reform process effectively through the key aspects of digital transformation and integration of shared services including the institutional set up of the reform process; mapping, capacitating, and redeploying of human resources, financial management; harmonisation of existing technology efforts and inter-ministerial coordination.

- Merging the management of civil service technology departments, including NITDA, Galaxy Backbone and other technology-related agencies into a central information and communication technology led agency tasked with providing and managing technology shared services, which will be shared by decentralised information technology support groups within the various MDAs of government.

- Reviewing of legal frameworks, statute, identifying legislative loopholes and facilitating the creation of the necessary legislative framework for national digital service enablement.

- Documenting the reform process for international and national knowledge sharing and learning.

- Reinforcing data and digital standards that should strengthen public data infrastructure. Data should now be considered as an infrastructure. It will underpin transparency, accountability, public services, business innovation and civil society.

- Investing in deepening the digital and data capabilities of public servants, this can be done by using pilot projects to develop an understanding of technologies (especially with reference to blockchain technology), together with the skills required to master them. The government can also consider the value of convening people from sectors such as finance, agriculture, ICT and healthcare to understand the common challenges each sector faces and then work together to address them. Enlisting this support will help maximise collaboration across society towards the common digital objective.

- Creation of a "*cross-government community of interest, bringing together the analytical and policy communities.*"

It can be said to be in the form of building a culture of constant change- making a change isn't about sitting back and waiting for the next five years of business as usual, rather it is about building a new momentum and rhythm in the government and public service which reflects the new reality of the industry.

DIGITAL TRANSFORMATION, WOMEN'S RIGHTS, POVERTY AND INCLUSION

Digital public service delivery and innovation can provide significant opportunities to transform public administration into an instrument of sustainable development. By reducing information costs, digital technologies greatly lower the cost of economic and social transactions for citizens and the public sector.

Digitalisation of public sector services could bring major benefits to the government in terms of reduced costs and engagement with citizens. Such can have a positive impact on a country's economy, competitiveness, and innovation. Accenture, a global management consulting, technology services and the outsourcing company says that a digitally enabled government could reduce back-office costs up to 45%, while a 1% increase in digitalization can mean a 0.5% gain in Gross Domestic Product (GDP) and a 1.9% gain in international trade. A 10% increase in digitalization could translate to a 0.86% drop in a country's unemployment rate. A digitally enabled government could also reduce back-office costs by up to 45% in their recent publication on digital transformation [72].

"The digitisation of nearly everything is poised to be one of the most consequential economic developments of the next ten years"
— John Chambers, Cisco Systems Chairman

Some of the immediate benefits of digital transformation that would have direct impact on poverty, marginalisation and women's rights are:

• Extending the reach of government services, making them more geographically, economically and socially available, and

192

- Promoting equal access to government services, thereby eliminating corruption and improving accountability which will lower opportunity for the elite to capture the process.

However, to achieve these benefits, a major requirement is the inclusion of women, girls, the handicapped, and other marginalised persons. This should be achieved by:

1. Defining the end-user: this must take into account the specific needs of women, girls and, for instance, people with disabilities, and people living in poverty.
2. Ensuring a pro-poor mandate: Ensure that all relevant institutions of government have a mandate and the financial and human resource capacity to represent the interests of the poor. Incentives may need to be provided to commercial service providers.
3. Setting relevant targets: for instance, pro-poor targets and the obligation to listen to citizens and encourage their participation need to be part of sector frameworks, and also part of performance and management contracts.
4. Inter-sectorial collaboration: no single actor can effectively deal with inequality on its own and collaboration with the social and welfare sectors, non-governmental organisations and individuals are required.

As the Nigerian government becomes more digital and ensures that majority of its citizens have the right access and digital skills, it will see much higher levels of engagement, accountability and public trust among its downtrodden.

DIGITAL TRANSFORMATION AND SUSTAINABILITY

On September 25, 2015, in New York, 17 Sustainable Development Goals were ratified as successors to the Millennium Development Goals which were launched during the Millennium Summit in 2000. Therefore, the MDGs are seen by many to have formed the basis for the new 17 Global Goals since many of its earlier goals and targets have yet to be achieved. The goals also have 169 ambitious sub-targets, which clearly define and describe the focus topics of each goal.

As ambitious as the goals are, public sector digitalisation can help significantly in fulfilling the majority of them. Since the key objective of delivering government services better and faster is the cornerstone of digitalisation within a fully interconnected network. Citizens suffering whether in the case of poverty, health care, gender equality, quality education, access to information, etc., are better served. Public service digital transformation will facilitate sustainable and resilient infrastructure development through enhanced financial, technological and technical support to all government Ministries, Departments and Agencies (MDAs) as seen in later chapters.

Digital Transformation promises the following actionable points towards sustainability in general terms:

1. Social Media: harness its cutting edge technologies, partners and approaches to maximise the impact of peoples' voice and perception of change.
2. Accountability and Monitoring: Empower ordinary citizens in every part of the country via both social and mainstream media, for accountability reporting of key government actions, policies, and projects.
3. Teamwork: Sponsor national and local citizen-driven accountability exercises via workshops, crowdsourcing and grassroots mobilisation, and rewarding/amplifying those with exceptional outcomes.

4. Reporting: Reporting to stakeholders in a transparent and public manner is fundamental to the success of the digitalisation efforts as well those of sustainability. There are several powerful drivers of non-financial reporting that are pushing institutions to act responsibly and report on their actions and inactions. Non-financial information is informing the decision of consumers, local communities, civil society organisations that are all expecting greater transparency from the government and businesses.

5. Access: The democratisation of public information empowers the less privilege to be able to press for their rights especially in worsening economic climate.

Figure 24: Sustainable Development Goals

The enormous challenges posed by the Sustainable Development Goals demand strong cooperation and coordination among all MDAs of government, together with the provision of adequate financing including private sector contributions and support.

DIGITAL TRANSFORMATION AND LOCAL CONTENT DEVELOPMENT

Nigeria has continued to receive hundreds of billions of dollars in the forms of rents, royalties and taxes from its primary export commodity – oil but with little impact on the lives of its people since the 1970s. For the above reason, there has been growing recognition that commodity export alone cannot fuel economic development, but broader policies are needed to foster diversification. This links and spills over development effects onto the local economy. Local content development and promotion has integrated itself into the Nigerian oil and gas industry in ways that a decade ago would have seemed unlikely due to a number of policy changes and implementation. Public sector digital transformation local content policy should require all participants whether local or foreign to source certain percentage of their intermediate goods or services from local producers and or sources. The policy should envisage a gradual increase of the percentage of inputs that needs to be sourced locally over time. The core objective of local content development in the digitalisation process is to develop and support local manufacturing and quality digital service provision through backwards, forward and sideways integration along its entire value chain.

According to Deloitte digital survey of 2015 [73], if government transformation through the power of digital technologies is a journey, then the inclusion of forward-looking local content policy must centre on infrastructure and human capital development. The policy must stress overall accumulation of physical and human capital as fundamental for economic development against trade. Of course, digital transformation is not just about technology but also about policy changes that can assist in the rejuvenation of local industries by focusing on encouraging local production and consumption amongst the people as against current practices.

Case Study: Trade Finance

Digitization and automation of trade processes has been on-going for many years, but the banks' updated processes are still largely based around the logistics of handling physical paper and documents. A lot of processes share similar characteristics, but requires completely different IT enterprise systems and procedural steps to manage. An example here would be documentary collection, letter of credit and consignment. All of these processes follow roughly the same five steps:

1.   Extension of credit to customer
2.   Informing the customer of credit status
3.   Banks open a communication channel regarding the customer
4.   Updating the status of goods from freight forwarder
5.   Execution of full or partial payment of funds based on certain criteria.

Blockchain technology could bring the benefits of automation to these trades. Through the use of cryptographic keys and multi-signature wallets, one can create a replacement for traditional trade finance documents, which are stored on the blockchain as a smart contract. The document is updated automatically by blockchain transactions as it moves through the steps of the trade process.

Two cryptographic keys are used to sign a blockchain transaction, one private and one public. This is analogous to how a signature on a physical document or a cheque proves validity in the physical world, only with an added layer of security. Ownership of the private key provides access to the digital assets stored at an 'address', which is analogous to accessing a bank account using a personal pin or passcode. Both keys are needed to create a transaction output, which means transferring your assets to another 'address'. Wallets are a service which do not store the assets themselves but they can generate, manage and store the needed cryptographic keys. Added functionality enables a wallet to require multiple signatures before the digital assets in them can be accessed, meaning that multiple parties can be involved in the same trade process [41]

SECTION III

Implementation & Monitoring

*"The drop of rain makes a hole in the stone, not by
violence but by often falling" – Lucretius*

Chapter 11

......................◆◆◆......................

FINANCING NIGERIA'S PUBLIC SERVICE DIGITAL TRANSFORMATION

Achieving the vision of Digital Transformation of Nigerian Public Services will require the mobilisation of significant financial and human resources. Making an authoritative budget is not easy. Verifiable and comparable data are difficult to come by. While it may be tempting to focus only on the benefits of digital transformation, the issue of project failure and unintended outcomes caused by inadequate or misplaced funding has the potential to hinder, or terminate the process. Funding is a major concern for several governments to provide core citizen services – let alone investment in digital initiatives. In contrast, digital transformation is widely recognised as a route to substantial cost savings. A study found that 82% of organisations perceive digital technologies as an opportunity, but only 44% managed to increase investment in those initiatives over the previous year [96].

On the other hand, many public bodies that started digitising to improve services are now placing equal emphasis on cost reduction. So, it is about balancing new revenue sources to finance transformation while exploring ways to reduce digital costs. For instance, through shared service arrangements, collaborative procurement and better management of contracts and use of cloud and open source systems. This approach will enable faster, more flexible and economic systems, development and maintenance, although issues of data security and privacy need to be carefully considered. Moreover, further ahead in digital transformation journey, adoption of advanced technologies and approaches, including mobile devices, big data, social media and data analytics, combined with changes in working methods, would allow local public services to be radically reshaped in ways that reduce costs and provide more streamlined and transparent services for citizens[96].

Basic components of costs, funding sources and leadership and standards required for digital transformation of public services are highlighted in this chapter.

BASIC COMPONENTS OF DIGITAL TRANSFORMATION EXPENDITURE

The overall costs involved in public service digital transformation can be classified as tangible and hidden costs. Hardware, software, network, integration, support and development costs are visible costs. While costs involved in organisation restructuring, change and change management and human reorientation can be classified as intangible costs.

Table 2 provides a list of these costs.

Table 2: Cost Components Involved in Digital Transformation of Public Services

Tangible Costs	Hidden Costs
Hardware acquisition, development and maintenance	Organization restructuring
Software acquisition, development and maintenance	Change and change management
Telecom networks	Human resource reorientation
Integration and process harmonisation	
User service provision (call centres, help desks)	
Performance management and quality control	
Research and development	

FUNDING SOURCES FOR DIGITIZING PUBLIC SERVICES

Digital technology is essential for Nigeria's economic diversification. The transformation will not only provide quick, convenient and efficient public services; it will also help drive growth in promising non-oil sectors and reduce reliance on oil-based incomes. However, financing digital transformation becomes critical considering tighter fiscal pressure, particularly when oil prices are lower. In such scenario, the government can explore multiple options to prioritise and fund digital initiatives to drive the transformation of public services. It can restructure budgeting and investments approach, reinvest savings from digital initiatives, leverage public-private partnerships and lease or rent capital assets among other income sources.

Restructure Budgeting and Investment Approach

Refining budgeting process will enable federal authorities to make more informed choices about where and how to reduce spending within and across MDAs as well as optimise its capital allocation and management processes. The narrowing tax-gap – difference between total tax due and actual tax receipt each year, capturing value from under-exploited federal assets and extending services charges, all can potentially increase cash-flow to the government treasury.

Output-based budgeting approach allows the government to determine to spend by looking at resulting outputs rather than budget at historical costs. By understanding the relationship between the maintenance budget, maintenance plan and outputs, the highways agency in Sweden determined that a different maintenance plan could achieve the same output and performance for a lower budget [4]. Moreover, optimising the management of tax collection, raising commercial fees where feasible, addressing fraudulent payments and pursuing stringent collection of non-tax receivables would result in incremental income to sponsor digitalization.

Reinvest Savings Resulted Due to Implementation of Digital Services

As public expenditure remains under pressure, a business case that describes the returns on digital investment is a key driving factor. According to government estimates, during 2014-15, United Kingdom government saved £1.7 billion through digital and technology transformation, including gov.uk savings of £61.5 million. The combined total savings over three years amounted to £3.56 billion [97]. Savings realised from digital initiatives can be reinvested to scale particular initiative further or to launch other digital services.

Leverage Public-Private Partnerships Innovatively

A joint study by McKinsey and Oxford University found that public-sector IT projects requiring business change were six times more likely to experience cost overruns and 20% more likely to run over schedule than such projects in the private sector [98]. Employing innovative arrangements with private sector players for service provisions also helps in cost management.

For instance, outcome-based payment mechanism should be adopted; wherein service providers are paid based on successful outcomes rather than the provision of specific activities. United Kingdom government have been piloting such arrangements in several areas such as children's services centres and prison rehabilitation and resettlement services, along with a small number of social impact bonds. Challenge.gov, a procurement portal set up by the US Federal Government's General Services Administration is another example of how governments can use success-based payments to spur innovation and reduce costs. The portal host's competitive challenges on behalf of more than 50 federal departments and agencies. Participants submit their solutions online, and the agencies benefit from all the solutions submitted, paying a prize for the winner. To date, the portal has run more than 380 competitions, received 42,000 solutions and dispensed $72 million in prizes [4].

Lease or rent Government-Owned Capital Assets

The government can undertake a number of schemes to monetize federal assets to fund digital initiatives. By leasing government premises, completely or partially, to private service providers, government authorities can innovatively connect rural and remote areas with digital technologies. For instance, Botswana and Bangladesh adopted similar 'rental solution' in rural areas. As one of the world's most sparsely populated countries, with rural electrification rates of around 10% and among the world's highest mobile and fixed broadband tariffs, the digital divide between city and village is stark in Botswana. Only 9% of Batswana use the internet regularly, and the majority are urbanites. The Botswana Government, in a public-private partnership with two of the country's three mobile operators, Mascom and BTC, created the Kitsong centre program to bridge the digital divide. Rather than building base transceiver stations or laying more fibre optic cable, Kitsong centres, which act as information offices for rural villagers and businesses, are leased to local entrepreneurs on a franchise basis. Run out of 197 Botswana Post offices; Kitsong centres provide villagers with access to computers, fax machines, and the internet. Batswana are also able to engage with their government in a more transparent and efficient fashion. The facilities can handle passport applications, and birth certificate and school registrations [99]

Similarly, other funding sources that government can exploit are vertical funding – development financing mechanism confined to single development domains with mixed funding sources; horizontal inter-institutional funding including the redirection of line ministry human resource and current ICT budgets; additional central assistance from the central government to the states that include centralized ad-hoc and Innovation Funds; loans; user fees. External financing from international actors involved in financing ICT are potential sources of support including donor countries, multilateral development banks, the UN, and the private sector via foreign direct investment and private sector led foundations and initiatives.

LEADERSHIP AND STANDARDS TO GOVERN DIGITAL TRANSFORMATION

The exponential changes that drive digital transformation challenge the established models of leadership and governance. A survey reported that just 32% of respondents believe their leadership has sufficient skills for meeting the challenge of digitally transforming public services [100]. The government need to restructure the organisation, redesign regulatory norms, empower workforce and explore new ways of project management for the digital era.

For instance, regulatory and physical challenges in Nigeria hamper digital infrastructure in rural and semi-urban areas, causing a widening digital divide. While access to mobile and internet is increasing (as covered in earlier chapters), this is largely among wealthier users with multiple devices and SIM cards and is clustered in urban areas. Digital infrastructure – access to the internet, computing and mobile, lags in rural regions. Restructuring regulatory norms are one of the key steps that government authorities should consider on priority to drive digital service delivery. For instance, recent imposition of the quality of service requirements (less than 1% call drop) discourages telecom operators to expand to rural and low-income areas in Nigeria. Similarly, central bank's restrictions on telecommunication companies from taking a mobile money operator role have escalated overall transaction fees, affecting adoption of digital services in rural regions[99].

Successful organisations in both the public and private sectors are increasing their budget allocation towards services than isolated projects, as it is expected that service owners will both run the service and be accountable for improving it. This enables quicker innovation, reduces barriers to change and allows executive leaders to set the direction without being stuck into excessive governance.

Similarly, mutually agreed benchmark and reference point would provide public sector leaders with a powerful baseline to assist their decision-making. Moreover, locally and nationally recognised standards for data collection would allow for constructive comparisons among local agencies. Effective coordination of digital investments across federal departments streamlines spending drives synergy and support common standards.

Finding the right way to optimise the structure, scale and operating model of government without creating the disruption and uncertainty of constant organisational change is important both in achieving fiscal consolidation and in ensuring the effectiveness of public services. There is no single ideal operating model for central government or models for interaction between MDAs. For instance, in Denmark, many of the functions performed by the centre sit under the Ministry of Finance, including the agencies for modernising public administration, digitisation, governmental management and IT services. The government established IT Projektraad to coordinate large-scale IT projects better across the government and generate cost efficiencies. IT Projektraad, a digitisation council reporting to the Ministry of Finance, function as its central IT steering group. The agency's goal is to ensure that the benefits and gains targeted in a project's business case are realised. This has allowed it to apply a test-and-learn approach, using pilot projects to ensure investments are effective and then bringing lessons learned to other agencies. To that end, the digitisation agency requires government institutions to adhere to specific methodologies and guidelines when planning their IT investments. It also develops and shares best practices, conducts risk evaluations for projects over a certain cost threshold, participates in project reviews, and helps oversee the government's IT project pipeline. Such central oversight has helped the Danish government reduce unnecessary investments, enforce common standards, and build greater project synergies[98].

DIGITAL TRANSFORMATION LEADERSHIP AND GOVERNANCE STRUCTURE

Technology leadership in Nigeria public service today is too fragmented; it is recommended that all digital teams within the various MDAs be brought together under a Chief Digital Officer (CDO).

This person should have the controls and powers to gain absolute authority over the user experience across all government digital services and the power to direct all government information communication technology spend. The CDO should also have the controls and powers to direct set and enforce standards across government departments in areas already discussed in this work including:

- Technical Standards
- Content Standards
- Design Standards
- Process and Process Re-engineering Standards
- Customer Standards

Digital governance should not be left to chance. Ineffective governance creates waste and missed opportunities, making digital transformation riskier and costlier than it needs to be. Governance requires conscious design and engagement by the country's most senior public servants. No governance model is optimal for all countries, but the lack of governance is never an option. The right governance model provides appropriate levels of coordination and sharing for digital initiatives, in line with the country's structure, culture, and strategic priorities

The transformation governance mechanism proposed for implementation must centre around two key digital initiatives with two different impact strategies:

- Sharing: Local, state and federal governments must use common sets of capabilities and resources (including people and technology)
- Coordinating: Local, state and federal governments must synchronise and align their broad overall objectives especially as it concerns prioritisation, standards, policies and compliance, etc.

Therefore the proposed governance structure must consist of the following:

1. **Digital Public Service Federal Executive Oversight Committee** led by the President with members drawn from key ministries like Finance, Works & Housing, Information & Communication, Interior, etc.
2. **National Steering Committee** led by the CDO making investment decisions, prioritising resources, ratifying policies and standards
3. **National Innovation Committee** led by the Chief Enterprise Architect supported by the CDO identifying technology-enabled opportunities, proposing rules and standards around new technologies.
4. **State Steering and Innovation Committees** by the designated State Technology Representative as appointed by the Governor supported by select group of state MDAs leadership
5. **Digital Public Services Advisory Board** led by IT directors from affected MDAs with support from select new digital roles.

Table 3: Digital Transformation Leadership and Governance Structure

	Role in sharing and coordination	Typical benefits and challenges
Oversight Executive Committee	Digital Transformation Sponsor	Benefits: Executive ownership of delivered digital public services
National, State, or LGA Level Committees	National level committees aim for coordination Steering committees: Making investment decisions, prioritising resources, ratifying policies and standards. Innovation committees: Identifying technology-enabled opportunities, proposing rules and standards around new technologies. However, some of the decisions and policies adopted by the committees may mandate sharing of resources and capabilities.	Benefits: Digital standards and policies, consistency across digital initiatives, resource optimisation, adoption of new digital trends. Challenges: Additional mechanisms are often required to lead transformation or to enforce standards and policies.
Shared Digital Roles/Units	Sharing is the main objective of these units. Resources with specific skills are pooled together to develop digital services for all units in the company. Some coordination also comes naturally as the units develop technology standards and implement policies governing the use of their services. However, coordination of digital initiatives requires additional mechanisms.	Benefits: New digital skills, shared digital services, economies of scale. Challenges: Structure and positioning in the organisation, coordination difficulties with local unit leaders, the definition of the "service catalogue."
New Digital Roles	New digital roles drive the use of shared digital resources, such as helping local units to adopt firm-level solutions or use centralised resources. These roles also coordinate across different initiatives and organisational units.	Benefits: Relay the digital strategy, help to enforce national level policies, facilitate the adoption of shared capabilities by local units, facilitate cultural changes Challenges Challenges: Positioning in the organisation, relationship with the local units, building the networks of local champions.

LEAD EXECUTING AGENCY ROLE

To provide a central command and control point for the recommendations in this book, a Public Service Digital Services executing agency is proposed. Its role should include all the under-listed for effective and efficient management of the transformation process:

a. The agency shall be a secretary to the Executive Oversight Committee that shall be chaired by the President of the Federal Republic of Nigeria. This committee shall be created by an executive order of the President, amendment of existing IT national legal framework(s) and shall approve all flagship projects under different MDAs, monitor the status of implementation of the Digital Services Roll-out and resolve any inter-ministerial conflicts and challenges.

b. The agency shall act a clearinghouse for all national ICT efforts and advise the President, through the various committees on all Public Service Digital Transformation projects.

c. The CDO shall be the chief executive of the transformation executing agency.

d. The Chief Information Officers or Director of Information Technology of the different MDAs in shall constitute a working group under CDO to be called the National Level Steering Committee that shall be chaired by the CDO. The working group shall scrutinise all ICT budgets and approve them before they are included in the national budget.

e. A working group of State Executive for ICT shall also be chaired by the CDO of the agency and shall have the mandate to review all State ICT budgets and projects and to align them to national priority areas as defined and approved by the President.

f. The Executing Agency shall have the overall mandate of overseeing ICT staff development in the public service. This will be achieved through a working group on ICT Professional Development under the MDAs and drawn from representatives of private and public sector organisations involved in professional development of ICT in the country.

g. The agency shall play an ICT leadership role and limited operational role. Since the government is large, this structure could lead to having a large and bureaucratic organisation. The agency shall, therefore, be transformed into an organisation that only provides high-end ICT professional services to the various MDAs, State and local government institutions implementing the large flagship projects.

h. All end-user computing and ICT operational services shall be carried by independent ICT units in each of the MDAs. The key high-end ICT professionals shall need to be considered under a separate scheme of service, comparable to those of leading public service enterprises, to be able to attract and retain qualified talent.

i. The agency staff members shall have very detailed knowledge of ICT, and many will have post-graduate degrees in business, engineering, computer science or information systems.

Case Study: Digital Cash will Boost GDP growth

With the impending 'death of cash' and the rise of digital currencies (such as Bitcoin), there are strong arguments for central banks to start issuing "digital cash" – an electronic version of notes and coins.

The Bank of England has released new research suggesting that a central bank-issued digital currency could lead to an increase in gross domestic product. The conclusions were drawn from a working paper titled "The macroeconomics of central bank issued digital currencies" [78] published July, 2016 that examined how a central bank-backed digital currency could yield macroeconomic benefits, while providing banking regulators a clearer picture of the financial system.

"First, it leads to an increase in the steady-state level of GDP of almost 3%, due to reductions in real interest rates, in distortionary tax rates and in monetary transaction costs that are analogous to distortionary tax rates. Second, a CBDC regime can contribute to the stabilization of the business cycle, by giving policy-makers access to a second policy instrument that controls either the quantity or the price of CBDC in a countercyclical fashion."[78]

"Without deviation from the norm,
progress is not possible"
— *Frank Zappa*

Chapter 12

..............◆◈◆..............

IMPLEMENTING AND MANAGING THE DIGITAL TRANSFORMATION PROCESS

The actions below are outlined to include an indicative list of expected results, flagship projects and a set of activities to be implemented, as appropriate. The proposed activities and projects are not prescriptive, exhaustive or exclusive at this stage. Rather they are illustrative of the core activities and flagship digital projects that would need to be undertaken, as appropriate, to achieve the desired results by 2020. The action plan is meant to complement other relevant initiatives and plans, including the NSPSR, the vision for Nigeria 20:20:20 and the new ICT legal and policy framework.

Digital transformations entail changes in culture, processes and technology systems, which are more challenging to implement in the public sector than that in the private sector. However, a large number of successful government initiatives globally support that by learning from private-sector best practices it is possible to achieve wider and deeper public-sector digitalization. Drawing parallel with Schein and Lewin's change theory (unfreezing, cognitive restructuring and refreezing), the three-phase organisational change process is recommended - Plan, Innovate and Establish New Structures to achieve comprehensive public service digitisation in Nigeria (refer to Figure 25).

Figure 25: Implementation and Management of Nigerian Public Service Digital Transformation

PHASE I: PLAN

Before government authorities embark on digital transformation journey it must define a comprehensive digital strategy and implementation roadmap, form lead executing agency to manage transformation process and assess its digital capabilities for successful transformation.

Define Digital Strategy and Implementation Roadmap

Digital transformation is about more than just technology implementation. A survey by MIT SMR and Deloitte highlights strategy as the key driver in the digital arena [98]. An organisation's digital maturity is influenced, to a great degree, by its digital strategy. Among respondents from government agencies at the early stages of maturity, only 14% say that their organisations have a clear and coherent digital strategy. In the case of more digitally mature organisations, the number grows sixfold, to 86% [96].

Developing a clear strategy and commissioning implementation charter is the first step toward successful digital transformation. Next, plan and implement federal, state or local government inter-ministerial strategy and implementation roadmap for citizen-centred digitalization of public services and administration. Then, construct digital transformation governance, structure and culture frameworks. A clear visibility and stakeholders' management strategy need to be designed which shows how digital transformation will support overall objective and vision. Short, medium and long-term plans should be articulated in line with transformation vision. Finally, implementation roadmap should lead to digitalization action plan.

The potential impact of a digital strategy is largely determined by its intent and reach. Rajendra Kumar, the joint secretary at India's Department of Electronics and Information Technology, says about the agency's key initiative: "*The goal of the Digital India strategy is to transform the governance of the entire country through digital interventions. The idea is that every domain of government should be able to deploy and use digital technologies in a manner that can increase the service-level standards, improve interactions with citizens, and raise efficiency*" [101].

Table 4 highlights major focal areas under each phase with operational objectives and key performance indicators.

Create Lead Executing Agency

As discussed in earlier chapters, digital transformation brings in multiple execution risks and challenges. To overcome new digital challenges and address opportunities around it, government need to set-up a central-level lead agency around its digital initiatives. This agency interacts across MDAs and coordinates actions among them, thus increasing sharing and synergy to get more value from digital initiatives [60].

For instance, the UK government set-up '*Government Digital Service*' agency and charged it with overseeing the country's digital strategy and implementing the transformation of its service provisioning to what it described as "*digital by default*". Strong central leadership and implementation provided by this agency successfully resulted in providing citizens, businesses, and government users with accurate, streamlined, and comprehensive services [98].

To create this central executing agency, the government should start by instituting a temporary policy supporting the operation of this central unit. The agency should be authorised to manage key aspects of transformation process including the institutional set-up of the reform process, mapping, capacitating/redeploying human resource, financial management, and inter-ministerial coordination by following time bound approach. Also, public service leaders should be empowered and incentivized to work across ministerial boundaries. Also, they should be encouraged to explore innovative approaches with a *"right to challenge"* whereby public organisations could apply for an exemption from an existing rule or regulation if they can demonstrate how they would be able to innovate and deliver better outcomes. To support open collaboration, standardisation and data sharing, shared service policies should be instituted for ICTs.

Assess and Build Digital Capabilities

Nationwide capability assessment checks maturity of technology architecture for e-governance, availability and quality of public service data, innovation potential and overall culture and readiness of ministries, departments and agencies (MDAs) for change. Digital talent is a critical element. Government authorities should evaluate its pool of digitally skilled employees, including competencies in programming, mobile, project implementation, digital marketing, social media and data analytics [102]. Audit of current-state ICT and other assets and liabilities of all MDAs need to be carried out.

Further, it is important to define and benchmark required-state needs to identify gap with current-state. For example, mapping of digital services, defining open data delivery standards and modelling public service delivery as desired in required state. Digital skills gaps need to be addressed to build capabilities supporting the digital transition. Top-cadre officials in each member states can be made transformation ready by developing their digital competencies through deep-dive immersion program (including technical, digital and management skills), connecting them to learning and sharing networks and collaborating with the private sector on best practices.

PHASE II: INNOVATE

The next phase of digital transformation involves the development of centralised infrastructure for service delivery, the roll-out of digital services and establishment of shared services within all MDAs.

Develop Centralized Infrastructure for Service Delivery

Legacy infrastructure hampers the progress of digital transformation. The government should provide initial capacity building and facilitate the establishment of functioning digital service architecture. New apps, databases, analytics and intelligence, need to be put in place to deliver near-real-time insights into citizen needs and behaviours. This will result in hybrid IT environment – supporting and maintaining existing applications and infrastructure while some elements may be retiring or replacing. Next, establish a centralised and shared common service delivery infrastructure. Also, establish a common technical infrastructure operation and support model to promote inter-ministerial collaboration. Open data initiative across all MDAs will drive transparency, citizen engagement and reduce corrupt practice in public services. For instance, in South Korea, under the "National Action Plan for Open Government Partnership," the Government is using digital to provide customised services to target groups of citizens. This program aims to identify new services the public needs [103]. Further, aligning organisation culture around digital, developing a digital marketing strategy, team structure and role definitions are essential for citizen-centred digital public services.

The Dutch government launched a comprehensive digital infrastructure project led by the National digital-governance agency, Logius. The project steering group included central and local governments and public IT agencies. Together, using world-class standards, they defined the technical specifications for the 13 central databases involved and their interconnections. They also created a government-wide dashboard to highlight project status and risks and used conferences and social media to disseminate and refine key lessons with public-sector IT managers around the country. As a result of these initiatives, physical visits to municipalities and government offices decreased significantly. For example, the number of visits to Rotterdam municipality offices decreased by around 50% from 2010 to 2013 [98]

Roll-Out Digital Services

Once necessary technology, infrastructure and skills capabilities are established, launch the pilot of digital public services for a local province. Carry a coherent roll-out of digital public services at national and state levels based on effectiveness and efficiency achieved from pilot initiatives. Commence digital public service marketing and citizen boarding process based on learning and outcomes from the pilot project. Adopt 'try and learn' approach to optimise systems and solutions and improve the success of digitising public service.

To put the ambition of public service digital transformation into action, the government can select some digital service projects based on priority, ease of implementation, impact on citizens and budgetary considerations. As an initial implementation, it can begin with digitising federal public service data. Then, it can be further expanded across all states and local authorities. Next, critical digital skills enhancement mission can be launched – starting with training public service employees and later scaling up for citizens. A single-point central – Digital Skills Development Authority should be established, which is empowered for assessment, promotion, development, coordination and governance of digital skills across the nation. Further, service provisioning at local agencies such as birth and death data, local tax payments, etc. can be selected for digitalization. Citizens would quickly avail these services at local service centres or some of them just by login to local body portal.

Table 4 describes selected digital public services that federal government can initiate. Considering internet penetration, use of mobile phones and social media in Nigeria, the governments can also innovatively crowdsource ideas from citizens on which public services to prioritise for implementation. This will encourage public participation and help in quickly gaining citizens' acceptance while rolling out these projects on the ground.

Table 4: Selected Digital Public Services, their Priorities, Goals and Expected Deliverables

Digital Service Projects	Selection Basis	Priority	Goals	Expected Outcomes and Deliverables
Affordable and Quality Internet Broadband	Aggressively expand network connectivity within the country (fibre optic ring across the country)	Very High	• WAN/LANs within and between government buildings and institutions across the country • Hotspots in public areas – bus stops, town halls, schools, etc. • Create more Internet Exchange points nation-wide	• Extension of ICT infrastructure to all National, State and Local Governments • Extension of quality and affordable broadband infrastructure to under-served areas around the country.
Nigerian Digital Currency	Development of a national digital currency	Very High	• Leverage the digitalization process, blockchain through homegrown digital currency development	• Domesticate digital currency operations in Nigeria with the support of the Nigerian Central Bank
National Public Key Infrastructure	Development of a secured and future proof PKI for Nigeria digital future.	Very High	• Leverage the digitalization process through homegrown PKI	• Easy access and verification of identities • Increased online transactions security
Create national middleware platform	Facilitate fast and effective access to/sharing of national public data repositories between public and private sector entities in-country.	Very High	• To support open government initiatives as already discussed • Leverage government data	• Increase economy performance of government data • Increase access, visibility and transparency
Integrated Persons Registry as part of a comprehensive Nigeria Persons Data Hub	Persons unique identifier based on a single registration system	High	• Development of persons related national data hub • Implementation of unique person identification system supporting online and offline integration	• Smart Identity cards for all Nigerians • Citizens' portal and authentication point. • Other system integration points- driver's license, passport, etc.
Integrated Companies Registry as part of a comprehensive Nigeria Company Data Hub	Company unique identifier based on a single registration system	High	• Development of companies related national data hub • Implementation of unique company identification system supporting online and offline integration	• Smart company identification system • Nigerian company's' portal and authentication centre.
Digitise Internal Federal / State Public Service Data under One Domain	Digitization of federal public service data forms the basis of further transformation process and easy to kick-start as control mechanism are internal	High	• To completely digitise federal data from paper-based records • To integrate information from all MDAs to one central domain/portal	• 'One-stop' online access to all federal public service data with built-in security and privacy factors

Digital Service Projects	Selection Basis	Priority	Goals	Expected Outcomes and Deliverables
Digital Skills Provision and Enhancement	Education around digital skills is necessary to make federal employees aware and prepare them for transformation journey. Awareness about digital technologies and their benefits to citizens during initial stages will increase chances of success of digital transformation.	High	• To train employees on digital skills who can then create, manage and improve digital services delivery • To educate citizens about digital skills and services, and empower youths for digital age jobs • To design and upgrade digital skills training curriculum and online portal	• Employees and citizens geared for provision and uptake of government digital services • Free online learning platform and push notifications across digital subjects • Comprehensive and on-going skills training programs to stay up-to-date with ever-evolving digital landscape
Digitise Local / Municipal Services	Citizens on broader level will be able to experience the results of digitising day-to-day public services early-on	Medium	• To enable systems, people and infrastructure at local levels to provide quick services to citizens	• Set-up of local service centres and local body portals where people can avail public services
School/ Community Network	Deploy homegrown integrated education management system to all schools	Medium	• Provision of education data, information, reports and statistics through the education portal	• Education Portal • School/Community Hotspots
Create ICT Centres of Excellence	Attract best brains and start-ups through funded CoEs	Medium	• Enact appropriate legislation to support start-ups, ICT CoEs, and Research centres	• Specialised CoEs development • Start-up development • Increased economic outputs.
Government Procurement Portal	Deploy an online procurement management system for routine operational procurement(later to cover all government purchases)	Medium	• A single point of access to all government procurement information, advertisement and bidding. • Integration of asset acquisition to a central asset data hub	• Reduce procurement cycle time and cost • Spread the wealth more evenly • Increase competition • Increase value and measure local content/ participation
Scaling up ICT Innovation	Develop programs to support commercialization of ICT innovation	Medium	• Science and Technology Parks nationwide • Technology Hubs	• Establish science and technology parks in all state of the federation • Implement support program for rapid commercialization of ICT innovations
National Assets Data Hub	Development of a National Asset data hub supported by assets management system	Medium	• Secure and increase visibility to national assets	• Smart identities system for addressing national assets • Transport information management system • Asset management system
ICT Human Capital and Workforce empowerment	Training and Re-training of university graduate to align their skills with industry requirements	Medium	• To attract quality universities and private sector participation in the re-training of university graduates for employment	• Develop MOOC ICT continuous education courses for training of trainers and the general public

Establish Shared and Common Services within all MDAs

Benefits of digital transformation will not be achieved when MDAs conduct digital activities in silos. Compartmentalization must be broken down in favour of open communication and cooperation. This will result in consistent access to information and services for citizens across government departments.

The platform-based approach to digital government shares technology services across departments and also procurement. This helps to reduce IT bureaucracy and duplication significantly across government and improve cost efficiencies. Close coordination of government agencies' web portal development, supporting infrastructure, cost control, and performance monitoring and evaluation are essential for this platform-based model to function. For instance, in Canada, national, provincial and local governments have developed a centralised, network approach to delivering integrated digital services to citizens. Governments have been successful in sharing costs and prioritising a common framework for technology infrastructure, identity management and leadership collaboration [104].

Consolidating and streamlining of national ICT delivery approach across MDAs will lead to better control and lower costs of service delivery. At a recent NITEC 2016 conference, Kola Aina, Group CEO of a Nigerian technology company said that "*So far there are pockets of successes (of leveraging technology by government) here and there, but it is not just about deploying computers in offices or every agency having its systems, it is really about creating linkages looking at how to have more shared services*"[105].

PHASE III: ESTABLISH NEW STRUCTURES

The concluding phase of digital transformation involves the building of governance and legislative frameworks and setting-up of knowledge management and training programs.

Build Governance and Legislative Framework

As digital tools and platforms are leveraged to provide better public services and drive citizen engagement, it becomes more important than ever to protect public data while optimising its use. It presents the policy challenge to balance the need to promote trust, transparency and accountability on user information, against user experience and open development of digital solutions. Thus, it is required to build executive leadership and governance structure supporting the transformation program.

Moreover, the governance of open, digital platforms must integrate required privacy, security and data protection throughout government MDAs. The whole-of-government privacy and security strategy in New Zealand is one of best-practice example. Here, a central governance group, including a Chief Privacy Officer, is charged with continually improving the privacy and security of state services, including tools, advice and guidance, as well as developing standards and assessments[106]. It is also important to identify policy and legislative gaps as well as facilitate the creation of a necessary enabling framework for public service digital transformation.

Set-up Knowledge Management and Training Programs

Knowledge-sharing platforms and communities can help transfer digital know-how across departments. This can include online portals of case studies and best practices, to disseminate learning from early adopters, as well as online communities and coordinated training and development specific to digital transformation (such as for the development of mobile applications). The Digital Services Innovation Centre in the US is designed to perform these functions under the US digital strategy, including through sharing solutions and training for digital services and infrastructure, such as user experience and security architecture [107].

Collaboration among employees will also help spread digital best practices and learning across all MDAs. The Chief Digital Officer (CDO) can also establish international relationships to collaborate on best-practice learning. For instance, the creation of "D5" group of digital-driven governments aims to strengthen the links between the public and private sectors among alliance member countries in technology and the digital economy. Members include the United Kingdom, Estonia, South Korea, Israel and New Zealand, with expansion expected in the coming years [108].

Table 5: Nigerian Public Services' Digital Transformation Focal Areas, Objectives and KPIs

Focal Areas	Operational Objectives	Key Performance Indicators
Phase I: Define Digital Strategy and Transformation Implementation Roadmap	• To develop and obtain stakeholders' acceptance of strategy charter • To develop a national, state or local government digital transformation strategy and implementation roadmap blueprint • To develop a national, state or local government visibility and stakeholders management blueprint • To develop a national, state or local government digital governance model supported by organisational structure emphasising a new digital culture – sharing, transparency, collaboration, etc. • To create the office of the nation's Chief Digital Officer under the office of the President of the Federation to lead the transformation effort • Appoint a national Chief Digital Officer "Transformer-in-Chief" to develop, nurture and implement the digital transformation exercise outside of the service with reporting responsibility to all stakeholders • Appoint digital champions across key MDAs to lead the transformation initiatives within their agencies and equip them with digital tools and skills	• Charter document commissioned and accepted by all stakeholders • National digital vision, mission and governance strategy developed and accepted • National Enterprise Architectural Model for digital service delivery framework developed and commissioned • Functional connectivity, central and shared national, state or local government data processing infrastructure platform implementation framework developed and accepted • Deliver strategy roadmap measurement criteria – success metrics, reporting and analytics, KPIs, etc. and available to the general public • Deliver a digitisation action plan

Focal Areas	Operational Objectives	Key Performance Indicators
Create Lead Executing Agency	• To create the Office of the Chief Enterprise Architect for the Federation under the office the Head of the Civil Service of the Federation • To create and legally capacitate a lead agency, the Digitization Secretariat, to steer the national digitalisation process under the leadership of the Chief Digital Officer of the Federation • To facilitate the lead agency's creation of an action plan for steering the national public service digitisation process • To facilitate the smooth transition of the agency to a central part of government for the management of public sector national shared services	• A legally capacitated lead agency mandated to steer the public sector digitisation process established. • A Chief Enterprise Architect appointed for the Civil Service Commission. • The lead agency is in procession of an action plan regarding its mandate
Assess Digital Capabilities	• To assess the capacity of civil service employees to administer and provide digital public services • To map the existing and required digitally enabled public services • To provide open data standards and framework for implementation • To provide a common operating model for all MDAs based on open, collaborative data sharing policies	• Digital capacity building needs assessment completed • Digital services skills development of public service officials completed • Mapping of existing public e-services completed • Digital service delivery model completed • Comprehensive report on ICT audit delivered
Phase II: Develop Centralized Infrastructure for Service Delivery	• To create the architecture for nationally relevant, citizen-centered digital public services and management • To commence limited pilot demonstration project around open procurement and collaboration – example, centralised email application management systems for all MDAs under a common platform • To commence digitisation marketing and awareness campaign	• A framework for digital service provision completed • A geographically limited pilot project carried out
Roll-Out Digital Services	• To roll out nationwide, the framework for digital public service provisioning system • To pilot a geographically unlimited digital public service provisioning system	• A framework for digital service provisioning and operation completed • A geographically limited pilot project reviewed, completed and lessons learned documented • Digital public service provisioning approach instituted nationwide
Establish Shared and Common Services	• To create a centralised shared services unit that supports a centralised system of digital public service delivery	• Legal and policy framework for central shared services unit completed • Operational mandate of central shared services unit defined and implemented • Merger of physical architecture and human resources into a central unit completed

Focal Areas	Operational Objectives	Key Performance Indicators
Phase III: Build Governance and Legislative Framework	• To create a governance structure supporting the transformation program's office directly by the President • To create the required policy and legal enablement framework for digital transformation, data privacy and openness in government • To complete the public sector institutional and policy reform required for digital public service enablement	• Legislation, policies and standards for digital public service delivery and administration • Public sector reforms which are organised to integrated digital public service delivery and administration
Set-up Knowledge Management and Training Programs	• To document the process of creating and implementing the digitisation process • To utilise various platforms for sharing the documentation and lessons learned • To develop multi-sectorial manpower development programs	• A knowledge management plan for the process completed • A set of knowledge sharing activities completed

Case Study: Broadband Effect on Increased Competitiveness

A number of examples show how broadband can help transform processes, business models, and relationships in the private sector.

A study involving business and technology decision makers from 1,200 companies in six Latin American countries (Argentina, Brazil, Chile, Colombia, Costa Rica, and Mexico) showed that broadband deployment was associated with considerable improvements in business organization, including speed and timing of business and process reengineering, process automation through network integration, and better data processing and diffusion of information and knowledge within organizations (Momentum Research Group 2005).

A 2005 study by McKinsey highlighted the growing importance of broadband to companies' competitiveness through new ways of structuring work. Modern companies build distinctive capabilities based on a mix of talent and technology; they specialize in core activities and outsource the rest. Broadband helps allocate activities more efficiently between workers tackling complex, highly dynamic tasks and more traditional, transactional workers. It is also a key component in raising the productivity of employees whose jobs cannot be automated, and in doing so cost-effectively (Johnson, Manyika, and Yee 2005).

Companies that adopt broadband and ICT to transform their supply chains prompt other companies in their value and distribution chains to adopt new technologies and interoperable IT systems (Atkinson and McKay 2007). For example, the automobile industry's entire technology-intensive supply chain is linked through broadband networks and high-power computing. Broadband networks are essential to engineering design service firms to test and implement design options directly with car and parts manufacturers.

Studies in the United Kingdom indicate that enterprises using broadband are more likely to have multiple business links. For example, they use e-mail and the Internet to raise the quality and lower the costs of gathering market intelligence and to communicate with suppliers and business partners. Enterprises with more links tend to have higher labour productivity [109]

*"It does not matter how slowly you go
as long as you do not stop"*
— *Confucius*

Chapter 13

·············•◆•···········

MEASURING THE DIGITAL TRANSFORMATION PROGRESS AND IMPACT

Digital technologies that are evolving as disruptive forces for public and private organisations presents serious roadblocks right from adopting new technologies to the reorientation of culture and quantification of its benefits. Connecting with citizens through digital channels, collecting feedback in real time, restructuring budgets, finding new revenue opportunities and in particular measuring and justifying the return on digital investments are some of the key barriers that make it difficult for governments and businesses to embark on a digital transformation journey. However, as described in earlier chapters, greater use of digital technologies will help Nigerian economy grow strongly and become more competitive.

Measuring Performance Indicators Helps Stay Aligned with Digitalization Goals

It all comes down to the way you measure success. The metrics you define to monitor how well your strategies are working will completely shape the kind of decisions you make. If you start by creating an intelligent and comprehensive measurement framework, it will focus your decisions and make you create better digitalization program.

Clarity in vision is required on how digitisation will be effective regarding quantifiable benefits. Lack of a clear yardstick to measure digitalization is an apparent impediment to successful transformation. It is critical to get a clear understanding about areas where digital initiatives will yield results, which may include operational efficiency, customer experience, reduced digital divide, etc. It is well-established that traditional KPIs are no longer effective at measuring the performance of digital initiatives.

For the assessment of digital government, individual indicators and composite indices have been developed by international organisations, academic establishments and individual countries. Some studies assess the use of ICT alone; others measure customer services through services offered by government websites. These services range from simple services to more sophisticated issues of privacy and electronic voting [110].

To measure and compare the digital transformation program, a set of feasible, relevant and comparable indicators are required. Two distinct sets of indicators are recommended in this book - the first being the digitalisation program itself which is to measure the programs and outputs. Second is to measure the outputs and impact of digital public services on the population over time.

For the output and impact of digital public services on citizens, a core set of indicators will be required. Methods may include traditional questionnaire-based surveys, administrative sources or collectable from country-level web surveys [111]. Whatever the data collection methods, digital transformation indicators should be statistically feasible, designed to enable international comparability, substantively relevant, consistent over time, understandable and accessible to policymakers and other data users and finally not so complex as to limit their collection and use.

Measurement of digital service performance helps to evaluate whether the service is meeting user needs, it allows users to easily complete the task; it is being used by enough people to make it cost-effective, and there is appropriate awareness about the service. Measurement of service should be initiated at the beginning of the program. Start by creating a performance framework, which outlines service objectives and what data should be collected to meet them. Then, check whether existing data analysis skills and tools available would be sufficient for the type and volume of data that will be generated from digital services. Explore how data will be gathered based on citizens' interactions with services and map how data will be stored, accessed and processed for insights. Figure 12.1 represents performance framework developed by Government Digital Service United Kingdom for transactional services such as registration, application and claims [113].

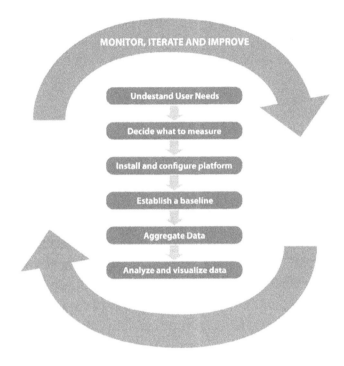

Figure 26: Performance Framework developed by Government Digital Service, United Kingdom

DIGITALIZATION PROGRAM PERFORMANCE INDICATORS

Benefits are realised when the outcomes are measured. It is critical to measure outcomes to ensure the digitalization program is driving improvement throughout the transformation journey. The indicators established during the implementation stage can be used in line with progress performance indicators to monitor and measure on-going progress. Monitoring program performance regularly can help in persistently fine-tuning or simplifying processes to generate more value from digital transformation.

Identifying clear set of indicators will also help federal authorities understand the status, trend and degree of success of digitalization program. This guides in better policymaking towards more efficient administration improved services and an equal participation for citizens. Table 6 represents selected indicators which can be used to measure the progress of digital transformation program.

Table 6: Digital Transformation Progress Indicators for Measurement

Progress Indicators	Description / Examples
Availability of digital technologies to federal organisations	Number of government offices having access to digital technologies
Use of digital tools by federal / State employees	Proportion of employees using computers, laptop, internet, required software, etc. in government offices
Service portfolio of various MDAs	Number of digital services offered, maturity, etc. (tax returns, registrations, permits, licenses)
Integration across MDAs	Collaboration and interoperability across government departments
Integrated payment facilities with digital services	e-payment / mobile wallet payment option with services
Development of digital skills	Number of trained digital champions in the public service
Cost per transaction	The cost to government each time someone completes the task your service provides is known as 'cost per transaction'
Public spending on digital initiatives	Proportion of GDP invested towards digital technologies/infrastructure/skill development
Initiatives to improve user experience of digital services[112]	
Multichannel provision	Availability of digital services on alternative channels – web, call-centre, email, mobile, social media, etc.
Privacy protection	Clarity on privacy regulation concerning personal data usage
Transparency of service delivery	Tracking and tracing of service provision, ability to conduct services in sequence and indication of time duration for service completion
Ease of use	Support in the form of FAQs, demo, live chat, etc. available
User satisfaction monitoring	Feedback options, and mechanism to monitor and manage user satisfaction

DIGITAL TRANSFORMATION IMPACT INDICATORS

Availability of digital services is only one step; however, government's focus must be placed on understanding and addressing the motives for use, so that adoption of services increases and citizens can reap actual benefits. For instance, the gap between the availability of services and their takeup shows that the public sector is facing important challenges to rethink how public services can become more citizencentric. Many countries now formally use methods for user needs identification and are moving away from a onesizefits all approach to digital services towards greater segmentation and personalization [112].

Table 7 represents selected indicators, which helps in determining the impact of digitising public services.

Table 7 Digital Transformation Impact Indicators for Measurement

Impact Indicators	Description / Example
Access to digital technologies to citizens	Number of active internet connections
	Number of unique mobile phone users
	Proportion of population with computer/laptop access
Take-up of digital services (adding further granularity – based on federal departments, geographies and demographic factors)	Use of digital services among citizens • Number of visitors to federal portals • Unique visitors • Number of registered users • Month on month growth in registrations • Number of downloads (forms, manuals, etc.)
Online / mobile-based payment transactions	Number of payment transactions carried out online / mobile-based while availing public services
Businesses related to digital technologies	Growth in number of ventures/businesses related to digital
FDI investment in digital space	Growth in foreign investments related to digital space
Public savings from digital services	Financial benefits generated by implementing digital public services
User review / citizens' trust in digital services	User satisfaction based on feedback mechanism • Number of comments / reviews on federal websites, emails, etc. • Photos / videos uploaded, shared and viewed • Sentiment of comments / feedback

Digital transformation needs clear goals and a way to measure success. However, it does not make sense to manage and measure transformation according to strict, short-term return on investment (ROI) metrics. Federal and state authorities need to move away from the traditional budgetary process and focus on the longer term. To truly leverage benefits of digital transformation program, the government need to aim and develop capabilities to track program performance and its impact in near real-time.

Case Study: Autonomous electronic Commerce and Intelligent Agents

Within a web of contextualized, blockchain-enabled smart contracts, it's conceivable that software agents could be set up to dynamically manage each distributed autonomous organization. In a virtually normalized data environment, those agents could reach out and grab knowledge and other digital or digitally represented assets.

One example of such a web would be autonomous e-commerce. A fleet of self-driving trucks delivers goods to distribution centers. The robots at the distribution center sort and move those goods onto autonomous delivery self-flying drones. Then the drones make deliveries to end consumers. An end consumer, meanwhile, has some returns to make (such as shoes that don't fit) and puts smart packages out for a drone to take back to the distribution center. Each step could be governed and executed by a web of smart contracts and one or more software agents acting as virtual third parties that have legal status [83].

Alternatively, consider what an intelligent agent will do for e-commerce?

- Imagine a system in which when a customer lifts the handset of a telephone to dial a long-distance international phone call, an agent within the smartphone automatically collects bids from the various carriers and decides which carrier to use. The telephone companies have an agent that automatically declares the price per minute for which it is willing to carry the call. The agent in the telephone decides which bid to accept, using an appropriate auction method (such as best bid wins and gets second best price).
- Imagine a system in which when a person wants to place an international phone call to a customer who does not speak the same language, an agent automatically translates the conversation between the two languages.

*"Whosoever desires constant success must change
his conduct with the times"*
— *Niccolo Machiavelli*

Chapter 14

....................◆◆◆....................

DIGITAL TRANSFORMATION KEY PERFORMANCE INDICATORS

As earlier discussed, the Fletcher School at Tufts University's Digital Evolution Index (DEI) 2015 performance score for Nigeria is 50th position in Africa in an assigned "*watch-out*" trajectory zone. The watch out zone comprises of countries facing significant opportunities and challenges, with low scores on both current level and upward motion of their DEI. Some of these countries such as Nigeria are considered able to overcome observed limitations with clever innovations, embarking on a digital transformation and other stopgap measures. But how should one describe the progress achieved in digitising public services? How will government officials get status on how Nigeria's transformation is progressing toward digital services? This requires a set of key performance indicators (KPI) that are specific to the digital transformation program.

'The Critical Few' concept by Avinash Kaushik refers to the imperative need to forget about the tonnes of meaningless metrics we have access to while keeping the focus on the few critical metrics for the transformation program. The problem that we face in the 21st century is not the lack of data, but its accessibility. Data creation continues to grow exponentially, and it is becoming more and more accessible as it grows. However, more data doesn't mean more intelligence. If one tries to measure too much, one will get distracted and lose sight of what's most important. However, if government organisation focuses on a few meaningful metrics, it will be able to stay focused on what matters most for the program, more effectively.

The exercise of defining success should produce a set of Key Performance Indicators. A KPI is a metric that can represent whether one is getting closer to achieving set goals or not. It's important that government organisations select only a few critical KPIs — i.e., choose only the metrics that are significant and relevant to its own objectives and discard all others. Then, based on these KPIs, the firm will be able to define relevant targets. For example, if one defines monthly visitors to the organisation's website as a KPI, in any given month, one could set a goal of reaching 150,000 visitors (this number becomes the target) [114].

DEFINING KEY PERFORMANCE INDICATORS FOR DIGITAL TRANSFORMATION

It is vital to set out a consistent way of measuring digital services performance. Because there will be no shared understanding of how concepts like 'citizens satisfaction' or even 'cost per transaction' are measured in government - which makes data-driven decision making difficult. For instance, Government Digital Service (GDS) in the United Kingdom wanted their service managers to be able to measure progress in three areas: improving the user's experience of the service, reducing running costs, and shifting people towards using the digital channel. They narrowed on the selection of KPIs based on design principles, which states that KPIs to be able to measure progress in three areas: improving the user's experience of the service, reducing running costs, and shifting people towards using the digital channel. GDS selected four KPIs – digital take-up, cost per transaction, user satisfaction and completion rate. To meet the service standard, all new and redesigned services must measure these four KPIs. But regarding performance measurement, service managers will add other KPIs to measure their more specific requirements [115].

It is recommended to start with defining key objectives of digital transformation (as described in the earlier chapter on 'Implementing digital transformation') and subsequent goals. Objectives would define strategic intentions of what government wants to achieve through digitisation program – question to ask: 'what is the purpose?' While goals are the specific strategies through which objectives are achieved – 'what should the outcome be?' Then KPIs are identified to measure achievement of goals – 'how do we measure the outcome?' KPIs are assigned targets, which are quantified or qualified expressions of objectives against which KPIs are measured – 'What are the quantified / qualified targets for each objective'. Based on the measurement mechanism, KPIs are measured, analysed, reported and required actions are taken [116] (refer Figure 27).

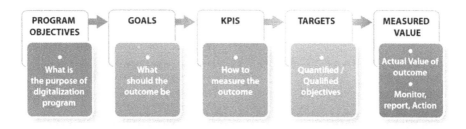

Figure 27: Key Performance Indicator Framework

For the planned digitalization program three groups of key performance indicators related to the output are recommended – infrastructure, governance and institution capacity and key impact indicators are discussed in this chapter.

Infrastructure: Review the Shift in Digital Capability Resulting from the Digital Transformation

Infrastructure-related indicators refer to the existence of techno-logical capabilities in the country that contributes to the success of digital services. It measures infrastructure readiness to achieve goals of digitalization. Improvement in connectivity, public spending on ICT capital and growth in the internet and mobile users are the few indicators that illustrate this group. Table 8 highlights output indica-tors under infrastructure group.

Table 8 Program Output Key Performance Indicators (Infrastructure)

Indicators	Description / Example
Improvement in Telecom, the Internet, mobile infrastructure in the country	Installation of telecom towers, roll-out of fibre-optic cables, better last-mile connectivity
Improvement in government data storage and processing capacity	Data storage space and data processing tools, analytics required by government
Increase in public spending on digital initiatives[117]	Proportion of GDP invested towards digital technologies/infrastructure/skill development
Growth in active internet users	Periodic (month/yearly) change in active number of internet users Average speed across the different regions
Growth in number of mobile subscribers	Periodic (month/yearly) change in active number of mobile connections Proportion of smart/feature phones
Improvement in Digital Evolution Index ranking	Change in digital performance scores given by Fletcher School at Tufts University
Integrated payment facilities with digital services	Facility of e-payment / mobile wallet payment option along with services
Data privacy protection measures	Appropriate measures to protect personal and transaction data of citizens

Governance: Evaluate the Change in Digital Governance Structure as an Effect of Digitalization

The governance structure is critical component supporting the transformation program. It is required to create policy and legal enablement framework for the program, data privacy and openness in government. Table 9 represents output indicators under govern-ance group.

Table 9 Program Output Key Performance Indicators (Governance)

Indicators	Description / Example
E-government management framework in place	Implementation of e-government plans
Digitization Secretariat in place and functioning	Appointment of 'Single point of contact' responsible for management of digitalization program
Transparency of service delivery	Tracking and tracing of service provision, ability to conduct services in sequence and indication of time duration for service completion
Existence and reforms of legislation and policies for digital public service delivery and administration	Additional policy measures / clarification declared by government to promote use of digital services
Effectiveness of framed policies delivery	Actual implementation and progress of policies launched
Contractual agreements for service provision	Agreements to share cost, risk and benefit and opportunity among different engaging parties

Institutional Capacity: Assess the Overall Services, Efficacy and Skills Enhancement of Organization Due to Transformation

Institutional capacity indicators are designed to assess the federal organisational and supporting systems readiness in digitising public services – which in turn leads to a change in performance. The geographical reach of digital services, digital skills development and progress in change management are some of the examples of institutional capacity KPIs.

Table 10 Program Output Key Performance Indicators (Institutional Capacity)

Indicators	Description / Example
Number of trained digital champions in the public service; size of digital workforce	Digital skills enhancement in federal departments; employee readiness for digital transformation
Availability of digital technologies to federal organisations	Number of government offices having access to digital technologies
Use of digital tools by federal employees	Proportion of employees using computers, laptop, internet, required software, etc. in government offices
Service portfolio of various MDAs	Number of digital services offered, maturity, etc. (tax returns, registrations, permits, licenses)
Progress in change management[118]	Managing transformation and re-engineering of processes and speed of access to information
Geographical reach of digital services	Launch of digital services by districts/towns/villages
Number of government systems operating at the defined service standards	Locally and nationally recognised standards allow easy comparisons among agencies

Indicators	Description / Example
Extent to which interactions are digital from end to end	Through digital service experience, rather than some physical form submission followed by online transaction

Program Impact: Measures Social, Environmental and Economic Outcomes of Digitalization

Key impact indicators refer to the desirable end-benefits of digital transformation. Social outcomes represent the impact that digital services may have on the society as a whole regarding participation, satisfaction and outreach. Wider reach and accessibility to citizens can also allow better ways of disseminating public information, thus improving communication between the government and citizens. Digital services will lead to reduced amount of paperwork, free-up citizens' time and provide electronic information transfer – supporting environment conservation. Economic outcomes relate to having greater opportunities of creating new jobs, new businesses and services – consequently boosting economic growth.

Table 11 Program Key Impact Indicators (Social, Environmental and Economic Outcomes)

Indicators	Description / Example
SOCIAL OUTCOMES	
Access to digital technologies to citizens	Number of active internet connections
	Number of unique mobile phone users
Centralized portal for digital services	'One-stop' online access/information of all federal public services
Citizen orientation to public services	User-needs focus approach in designing digital services
Digital services adoption by the low-income and rural population	Adoption would indicate wider and deeper reach of initiatives and will drive convenience in remote areas
Integration across MDAs	Collaboration and interoperability across government departments
Initiatives to improve user experience of digital services[112]	
Multichannel provision	Availability of digital services on alternative channels – web, call-centre, email, mobile, social media, etc.
Ease of use	Support in the form of FAQs, demo, live chat, etc. available

Indicators	Description / Example
User satisfaction monitoring	Feedback options, and mechanism to monitor and manage user satisfaction
User review / citizens' trust in digital services	User satisfaction based on feedback mechanism • Number of comments / reviews on federal websites, emails, etc. • Photos / videos uploaded, shared and viewed • Sentiment of comments / feedback
ENVIRONMENTAL OUTCOME	
Take-up of digital services (adding further granularity – based on federal departments, geographies and demographic factors)	Use of digital services among citizens • Number of visitors to federal portals • Unique visitors • Number of registered users • Number of downloads (forms, manuals, etc.) • Number of transactions (with and without payment)
ECONOMIC OUTCOMES	
Public savings from digital services	Financial benefits generated by implementing digital public services
Number of business establishments in digital sector	Government support (financial, technical or regulatory) to encourage new ventures/startups in digital space
FDI investment in digital space	Growth in foreign investments related to digital space

Implementing a new KPI measurement system appropriate for the digital age is challenging in both the public and private sectors. According to a survey, less than 40 percent of executives say their companies have set targets for their digital products or designated "owners" responsible for their success. Instead, they rely solely on the KPIs they have traditionally used, such as the number of problems resolved through call centres or the speed with which employees reply to written mail. Such traditional KPIs will remain important and should still be tracked, since many customers may prefer in-person, phone, or postal interactions, but they are no longer sufficient to evaluate customer service.

The experience of a government agency in Europe illustrates the problems that can occur when traditional KPIs are the only metrics tracked. The agency successfully implemented a new online system to promote customer collaboration but, in keeping with its traditional KPI system, only gave employees credit for face-to-face interactions. Even replies to e-mails were considered irrelevant. This practice inadvertently sent a signal to employees that the online channel was unimportant, and it failed to gain traction after a strong start. Thus, although governments should begin thinking about digital KPIs at the earliest stages of strategy development, the selection process requires long-term deliberation[98].

Case Study: The importance of a national Public Key Infrastructure (PKI)

Nigeria's' ability to participate within the global economy is seen to be dependent upon its ability to have a trusted, secure environment to conduct business in the electronic medium. A national (PKI) therefore needs to contribute solutions to the following issues:

- Privacy - keep information confidential;
- Access control - only allow selected recipients access to the information;
- Integrity - assurance that the information has not been altered;
- Authentication - proof of the originator of the information; and
- Non-repudiation - proof that information was sent by the originator.

The importance of a national PKI or a national electronic authentication framework cannot be over emphasized. Research continue to support the case for the establishment of a national body to oversee electronic authentication [79], the key justifications are:

- to promote compatibility;
- to present and represent a single national view;
- to ensure user confidence;
- to provide consumers with reliable information;
- to promote a contestable market for Nigerian based Certificate Authority services;
- to manage systemic risks;
- to facilitate the provision of value-added services (which rely on a robust PKI);
- to support legislation; and
- to promote export of trust-based services.

*"There is no room for complacency in
the fast-moving digital world"*
— Neelie Kroes

Chapter 15

·············•◆•·············

EVALUATION PARAMETERS AND REPORTING ON NIGERIA EXPERIENCE

This chapter discusses the reporting structure, combining the indicators into an annual presentation of goals, KPIs, gains, and next steps. Then the report on the importance of information dissemination is detailed followed by implementation schedule. The chapter ends with a description of the need of digital skills enhancement to maintain the progress of digitalization.

Reporting the Performance of Digitalization Program Helps Estimate the Benefits Achieved

When digitalization initiatives get steadily implemented and rightly measured at the same time, it becomes important to monitor their performance and report progress. The measured and reported data will help guide strategic decisions and pursue corrective measures. Moreover, government authorities and citizens demand to see the value achieved from digital investments. Analysis of reported data will assist in estimating benefits gained from digital transformation of public services.

For international reporting purposes, where possible, countries should provide numbers for indicators rather than proportions. This makes it clear what the data mean and facilitates comparison of data across countries. It also enables the aggregation of subcategories (for example, size categories). Population estimates for the total population, and for each subpopulation (as indicated by the classificatory variables), also needs to be provided so that proportions can be derived. Both sets of numbers should represent the whole population and not a sample [110].

Table 12 outlines a sample quarterly report to monitor and analyse the success of digital transformation based on measurement approach and key performance indicators recommended in previous chapters. Goals aligned KPIs are measured every quarter, compared to targeted values and accordingly next steps are planned to achieve desired objectives. In the initial stage of transformation roll-out, it is recommended to measure, monitor, report, and analyse KPIs every week, then shifting to monthly frequency and finally as a quarterly activity.

Table 12 Quarterly Report on Goals, KPIs, Gains and Actions Required in Achieving Public Service Digital Transformation (draft report outline, not exhaustive)

	Goals	Key Performance Indicators	Target Values* (Q1 Year 1)	Gains / Measured Values (Q1 Year 1)	Incremental Actions / Next Steps
SOCIAL OUTCOME	To increase citizen's access to basic infrastructure required for digital services To create centralised portal offering complete federal public services To integrate systems across MDAs To drive multichannel provisioning of services To monitor citizen's trust and satisfaction about digital services	Number of active internet connections	50%	__%	Increase broadband roll-out by --% in remote areas
		Number of unique mobile phone users (real subscriber base)	60 million	__ million	Drive penetration by --% in rural and areas
		'One-stop' online access/ information of all federal public services (percent achievement)	20%	__%	Integration of public service 1 and two by Q1 Year 1
		Collaboration and interoperability across government departments (per cent achievement)	30%	__%	Establishing messaging and cloud services for public service staff
		Multichannel availability of digital services (percent achievement)	35%	__%	Launch citizen supports helpdesk
		User satisfaction based on feedback (index from low 1 to high 10, based on comments, uploads, sentiment, etc.)	4	__	Address concerns highlighted by citizens to increase adoption and satisfaction
ENVIRONMENTAL OUTCOME	To drive take-up of digital services Add further granularity based on federal departments, geographies and demographic factors	Number of visitors to federal portals	5 million	__ million	Drive awareness about launch of digital services
		Number of unique visitors to federal portals	2 million	__ million	Increase promotion of availability of digital services
		Number of registered users	1 million	__	Build customization features such as auto-fill of forms based on registered user profile
		Number of downloads (forms, manual, etc.)	200,000	__	Upload more forms, documents and policies for citizens reference and use
		Number of transactions (license issuance, tax payment, etc.)	100,000	__	Leverage advocacy of registered users to drive adoption among non-users

	Goals	Key Performance Indicators	Target Values* (Q1 Year 1)	Gains / Measured Values (Q1 Year 1)	Incremental Actions / Next Steps
ECONOMIC OUTCOMES	To generate public savings from digital service. To foster conducive environment for business establishments. To drive FDI in digital sector	Total savings realised from digital transformation (against baseline year)	$300 million	__ million	Improve savings by increasing efficiency within existing digital services and implementing new services
		Number of businesses registered in digital sector (post roll-out of digital transformation)	10,000	__	Streamline processes to start a business, improve infrastructure and labour laws
		FDI in digital sector (post roll-out of digital transformation)	$200 million	__	Recommend financial, technical and regulatory reforms required to boost FDI inflows

Targets mentioned here are random and just indicative; it should be replaced with feasible numbers based on identified strategic objectives

Dissemination of Insights is Critical to Drive Agility in Decision-Making

Even if information is at the very centre of digital transformation, the link between information management and digital transformation is not made often enough. The four core challenges of information management highlighted are – How do we get any business insight out of all the information we are collecting? How do we use information to better engage customers, employees and partners? How do we manage the risk of growing volumes and complexity of content? And finally, how do we automate our business processes? [119] Working through and above silos making data, information and insight from analysis available when and where it matters are what truly makes the difference. It is required to build government-wide understanding and effective use and benefits of digital technologies, however having a culture of knowledge dissemination is also important.

Periodic progress report on digitalization progress is necessary, but report data in isolation are not sufficient to drive agility in decision-making and actions. It delivers value when it is analysed and then percolated across the organisation, in particular to a federal employee who need the insight most to act swiftly. Thus, it is recommended to include a governance process to prioritise report data, a high level of cross-functional coordination so that different parts of the organisation can actively listen and share what they know, and an IT infrastructure that facilitates the capture, analysis, and dissemination of relevant information. This will enable the federal workforce to make fact-based decisions promptly. Federal authorities need to completely re-imagine their underlying processes and build digital components to achieve maximum benefits of dissemination of information on digital transformation. Digital components required would be knowledge management systems to organise insights, collaboration systems to facilitate remote conversations, dashboards to display relevant information, and analytics systems to provide evidence-based insights to support decision making [120].

Digital Transformation Program Implementation Timeline is Projected to be 48 Months

The digital transformation journey is a temporary one, as eventually digital government becomes business as usual. So, KPIs for digital transformation are also provisional. Good KPIs should be time-bound; therefore, a timeline of program implementation is established.

How does an enterprise know when its digital business transformation is complete? According to Gartner, it is really up to the enterprise to define this endpoint. Although there can be on-going incremental improvement indefinitely, it is a good practice to set an endpoint to convey a sense of accomplishment. Endpoints will provide a strong sense of structure to the digital business transformation program — which executives will appreciate [121]. We recommend 48 months' timeline for digital transformation of public service in Nigeria. Figure 28 outlines implementation schedules based on eight focal areas as discussed in the previous chapter on implementing digital transformation.

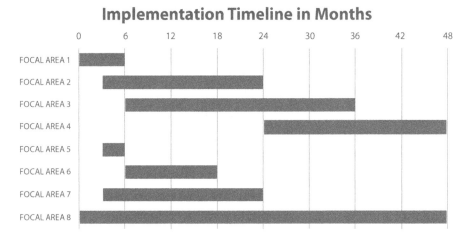

Figure 28: Timeline for Digital Transformation of Public Service in Nigeria

DIGITAL SKILLS ENHANCEMENT ACTS AS A KEY DRIVER OF DIGITAL TRANSFORMATION PROGRESS

Learning about digital skills and training needs from the Government Digital Services, UK

Skills shortages have been identified as key risk to digital government strategy, suggesting that federal authorities need to drive recruitment and capability building initiatives. GDS UK business plan for 2014-15, describes that "*we have difficulty hiring and keeping skilled staff. We will address this by making plans and processes to attract and keep the best digital and technology people; defining clear career paths; identifying development opportunities for staff; reviewing compensation and reward packages to meet market expectations [122]*".

GDS UK published '*The Civil Service Capabilities Plan*' and identified digital capability as a high priority. Under this plan, all departments will be using the same set of standards to recruit and promote staff and for the first time in 20 years, it will have a common approach to performance management. To improve Civil Service competencies, the plan identified four priority areas - leading and managing change, commercial skills and behaviours, program and project management, and redesigning services and delivering them digitally (digital skills). The plan is, at its core, about people and skills – how we train individuals and develop their competencies. It also considers organisational structures and management processes – how to structure, manage and deploy these skills to maximise people potential. The plan sets out what our leaders need to do, what needs to be done at a corporate level, what departments need to do and finally what individuals need to do to build their capabilities (see Figure 29)[123].

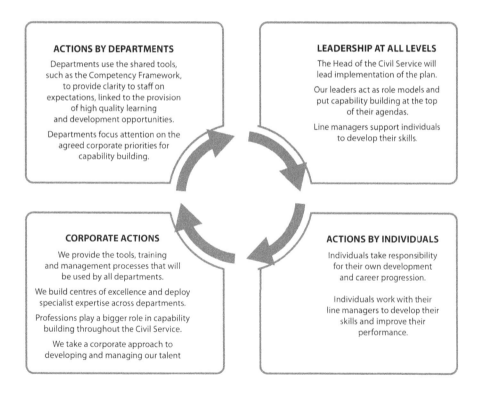

Figure 29: GDS UK – Civil Service Capabilities Plan [123]

The governments need to initiate training programs on digital tools, look at innovative recruitment methods, enter into partnerships and engage with the start-up community to develop digital skills. P&G is an interesting example of a company that has trained its employees on digital skills to scale up for future growth. Procter & Gamble established a baseline for all its employees through a "digital skills inventory," and set proficiency expectations for specific roles and career progression. To step up its Internet marketing initiatives and to scale up digital skills amongst its employees, P&G and Google started an employee exchange program. The aim of the program was to foster innovation and cross-pollination of digital talent. Employees from both companies took part in each other's training programs and attended meetings where business plans were formalised. With this program, P&G gained expertise in digital and search marketing to effectively sell its products online [124].

Once the methods to bridge the digital skills gap have been implemented, it is crucial to establish a continuous system of monitoring progress and regular upgrading. In the case of internal training, for instance, organisations could start by rolling out a basic training and evaluation system. Companies should continue evaluating their programs as before but use real-time technologies to increase their agility by adjusting training content and couple it with more frequent talent performance evaluation.

Developing the digital skills of federal workforce and citizens is not only a pre-requisite to the development of digitalization but a critical element in an increasingly digital world. It is important to ensure that digital skills enhancement, implementation and exploitation are an integral and continuous component of the development of digital transformation program. The government will have to continue working towards sustaining the advantage that they gain through these digital skills. This will require sustained efforts towards training and re-skilling.

Case Study: Central Public Procurement Portal

In pursuance of the recommendations of the Committee on Public Procurement headed by Shri Vinod Dhall and the decisions thereon of the Group of Ministers constituted to consider measures to tackle corruption and ensure transparency, the Department of Expenditure has taken action for setting up a Central Public Procurement Portal (CPP Portal). National Informatics Centre (NIC) has been given the responsibility for setting up this portal.

The primary objective of the portal is to provide a single point access to the information on procurements made across various Ministries and the line Departments. The CPP Portal is accessible at the URL http://eprocure.gov.in and it has e-publishing and e-procurement modules. It will be mandatory for all Ministries / Departments of the Central Government, Central Public Sector Enterprises (CPSEs) and Autonomous and Statutory Bodies to publish all their tender enquiries issued on or after April 2012.

The portal addresses the following salient points:

- Platform for publication of tender and bid award details by Central and State Government Ministries, Departments and Organizations,
- Free Access to tender documents for all
- Facility to publish all government RFQs through on-line form,
- Facility to publish tender documents, to add corrigendum with document,
- Facility to publish bid award details along with contract document and awards,
- In public domain, no registration required for access/viewing by public
- Email / SMS alerts at various milestones.
- Toll Free Help Line facility for all.
- Categorized according to types of tenders, product categories, types of organisations and organisation name
- Archived tenders are available in public domain
- Search facility available using parameters like keywords, location, value, organization name, organization type, tender type, and product categories.

The portal's main goals and guiding objectives are:

1. The prime objective is to educate, encourage/ incentivize and thus facilitate Government Departments at all levels, through appropriate steps and interventions from Central Government level, to readily adopt and use the Central Public Procurement Portal to bring in transparency in tendering process and gradually move towards adoption of electronic Procurement solution(s) for their procurement needs on a continuing basis.

2. Act as catalyst in streamlining the procurement processes of public/ Government sector with the help of ICT tools and technologies, enabling them to harness the multidimensional benefits of e-Procurement/ e-Tendering pertaining to: efficient and cost-effective procurement, shortened procurement cycles, full transparency in the whole process, avoidance of human discretion/ interference to the extent possible, easy availability of complete audit trail and evidential data etc.

3. Enable access to widest reach of tenders and unhindered secured bid submission facility for all, from any corner of the country [80].

"I will act as if what I do will make a difference"
— William James

Chapter 16

.............•◆•.............

CONCLUSION AND RECOMMENDATION

"We shall correct that which does not work and
improve that which does. We shall not stop, stand or
idle. We shall, if necessary crawl, walk and run to do
the job you have elected us to do."
— President Muhammadu Buhari

The President Muhammadu Buhari administration and leadership have an unprecedented opportunity for change and needs to pursue civil service digital transformation as a matter of urgency.

Here's why. Even in the midst of substantive reforms across the country, public debt is still rising; youth unemployment remains at disconcerting levels and overall health and Gross Domestic Product (GDP) growth severely strained, reflecting an underlying trend of a failing state. As government braces for these inconvenient truths, technological disruptions have fundamentally changed the way our people live, work, interact and learn. Private sector organisations have been quick to adapt. For instance, the retail, telecommunication, and banking industries are leading the way in fulfilling their customers' life needs through "always-on" digital channels.

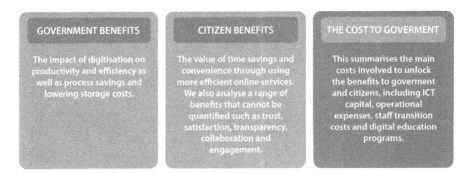

GOVERNMENT BENEFITS	CITIZEN BENEFITS	THE COST TO GOVERMENT
The impact of digitisation on productivity and efficiency as well as process savings and lowering storage costs.	The value of time savings and convenience through using more efficient online services. We also analyse a range of benefits that cannot be quantified such as trust, satisfaction, transparency, collaboration and engagement.	This summarises the main costs involved to unlock the benefits to goverment and citizens, including ICT capital, operational expenses, staff transition costs and digital education programs.

Figure 30: Digital Transformation Benefits

The reality is that these digital consumers are also digital citizens who have similar expectations of their government to provide dramatic changes in the way they operate and offer high-quality services while delivering value for money. There is a strong consensus on "what" needs to be done and "why." Themes of public trust and accountability, productivity, openness, innovation, co-creation, a new mindset for public leaders with much more "*digital adoption*" are generously peppered throughout this proposal. Overall, public trust has taken a beating through years of endemic corruption, inefficiency and poor service delivery by the Civil Service.

Evidence from many countries such as Estonia, the UK and other EU states, shows that the future of public service delivery is in digital transformation. Digital public service delivery and innovation can provide significant opportunities to transform public administration into an instrument of sustainable development. For a country with economic, social and political challenges – and opportunities – such as Nigeria, a digital transformation of the public service will enrich citizen-public service interactions and foster trust and ownership of the government.

Digital transformation will bring Nigeria's public service to the standards that have been set for service delivery across the world and will fulfil Nigeria's ambitions for its people and its public service. Public sector digital transformation when fully blended with the overarching public sector goals will enable the Nigerian public service to create an enabling environment for economic growth and innovation, to cut costs in its operations, reduce the amount of time spent accessing its services and dramatically reduce corruption and increase accountability.

Moreover, digital transformation holds the potential to contribute to reducing inequality, and would greatly enhance the productivity of women, girls, people with disabilities and other marginalised groups. To enable the marginalised to access public services with the same ease and ability as the more fortunate requires that these services are planned and implemented with knowledge of the diversity of the end users. Working with a diverse range of stakeholders will ensure that the digital transformation process is about the citizen and not about the provider.

For a nation that is composed mostly of young people, the concept that the public sector can be effective, noble, and helpful and that it can have the interests of its clients at its centre might be novel. This applies to the civil service itself. Can it conceive of its new digitally enabled, effective self? Changing the civil service culture will be a difficult undertaking, and one that will require a committed leadership, a leadership that will be able to visualise what is possible with digital transformation.

It is possible for the Nigerian Civil Servant to do more with less.

RECOMMENDATION

Governments, Nigeria included, in developing countries today are grappling with a myriad challenge of governing in very demanding times. Fewer budgetary resources, legacy overhead from longstanding institutional inefficiencies, limited access to training resources, a more demanding public now enamoured to the always-on, real-time online access to services and information. All these factors have worked to exacerbate this situation. As a result, the challenges of governance in a developing environment in transition have never been so great.

In all this turmoil, government officials are generally aware that there exist several systems and automated alternatives that specifically address all these challenges, in a cost effective and performance manner, but don't have accurate but accessible information about what specific solutions are available, how the various technology solutions address these type of burgeoning needs or what the best practice approaches are towards implementing these digital trans-formations.

The purpose of the Recommendation is to help governments adopt more strategic approaches for the use of a technology that spurs more open, participatory, and innovative governments. Key actors respon-sible for public sector modernisation at all levels of government (from co-ordinating units, sector ministries, and public agencies) will find the Recommendation relevant to establish more effective co-ordi-nation mechanisms, stronger capacities and framework conditions to improve digital technologies' effectiveness for delivering public value and strengthening citizen trust.

Many recommendations have been made throughout this book, however here a presentation of summary recommendations for Public Service Digital Transformation is presented with the aim of bringing governments closer to citizens and businesses:

i.   Governments must increase its investment in broadband connectivity across the country and unleash the power of the new digital economy by providing free and accessible broadband access to all Nigerians.

ii. Government must develop and implement digital strategies which:

   •   Ensures greater transparency, openness and is inclusive of government processes and operations.
   •   Encourage engagement and participation of public, private and civil society stakeholders in policy making and public service re-design and delivery.
   •   Create a data and technology-driven culture in the public sector.

iii. Government must update, build and adopt best practices in national information security procedures, policies and standards.

iv. Government must strengthen international cooperation with other governments to better serve citizens and businesses across borders, and maximise the benefits that can emerge from early knowledge sharing and coordination of digital strategies internationally.

In implementing the digital Government strategies, governments should:

1.   Develop clear and unambiguous digital public service business cases to sustain the funding and focus on digital technologies in the country, by:
   a.   articulating the value proposition for all projects above a certain budget threshold to identify expected economic, social and political benefits to justify public investments and to improve project management;
   b.   involving key stakeholders in the definition of the business case (including owners and users of final services, different levels of governments involved in or affected by the project,

and private sector or non-for-profit service providers) to ensure buy-in and distribution of realised benefits.

2. Ensure that general and sector-specific legal and regulatory frameworks allow digital opportunities to be seized, by:
   a. Reviewing and amend existing legal and regulatory frameworks as appropriate;
   b. Including assessment of the implications of new legislations on governments' digital needs as part of the regulatory impact assessment process.

3. Reinforce institutional capacities to manage and monitor projects' implementation, by:
   a. adopting structured approaches systematically, also for the management of risks, that include an increase of evidence and data captured in the course of project implementation and provision of incentives to augment data use to monitor projects performance;
   b. ensuring the availability at any time of a comprehensive picture of ongoing digital initiatives to avoid duplication of systems and datasets;
   c. establishing evaluation and measurement frameworks for projects' performance at all levels of government, and adopting and uniformly applying standards, guidelines, codes for procurement and compliance with interoperability frameworks, for regular reporting and conditional release of funding;
   d. reinforcing their public sector's digital and project management skills, mobilising collaborations and partnerships with private and non-governmental sector actors as necessary;
   e. conducting early sharing, testing and evaluation of prototypes with the involvement of expected end-users to allow adjustment and successful scaling of projects.
   f. Procure digital technologies using open procurement practices based on an assessment of existing assets.

Epilogue

It is not a secret that any African country must redouble efforts to build efficient, resilient and capable states. The modern way to improve the activities of public sector services is to use the Information and Communication Technologies (ICTs). There are many world-wide examples which allow the saying that the sooner Africa will have governments which are fully tech-enabled with a tech-savvy workforce is the better.

The World Economic Forum's Global Agenda Council on the Future of Government in its report "The Future of Government: Lessons Learned from around the World" (see http://www.weforum.org/reports/future-government) recommends "... flatter, agile, streamlined and tech-enabled (FAST) government". The Word Bank's report (see http://www.etransformAfrika.com) also presented the potentials of ICT to transform business and government in Africa. The UN global surveys (see http://unpan3.un.org/egovkb/global_reports/10report.htm and http://unpan3.un.org/egovkb/global_reports/12report.htm) have proved that the deployment of e-government brings the following advantages:

- public trust that is gained through transparency;
- better financial regulation and monitoring thus reducing in the possibilities for corruption;
- increase in the performance of governmental agencies;
- bridging the digital divide by reaching out to vulnerable populations.

Also, e-government is strongly associated with the country's level of democracy.

As it is necessary to build e-government for practically all 55 countries in Africa, it is logical to consider an opportunity to combine efforts from various national programs. But how create synergy between common tools, functionality & services and the vast diversity of countries their different political systems and unique challenges?

Right now, the modern ICT industry knows the answer to this question which is a digital platform approach. Considering that e-governments at the continent are not well-developed yet, a pan-African digital platform will be a green-field project which can be done correctly from the beginning thus saving money in the long-run. Such a platform immediately creates numerous opportunities for public-private partnership by facilitating provisioning of public, social, professional, voluntary, private and commercial services.

This is a unique opportunity, because of the combination of the following factors:

- overdue need for improving governance, transparency, performance, and traceability of the public sector;
- high economic effect of implementing once and re-use about 50 times;
- no need for huge up-front investment as e-government can be deployed incrementally with the pace of each African country;
- e-government is a green-field project which can be done with high level of quality at the entry point;
- higher level of e-government is associated with the higher level of democracy;
- real example of continental public goods;
- world-wide progress of ICT tools;
- advancement of the ICT infrastructure in Africa;
- entering of major IT vendors to Africa;
- unlocking potentials for the PPP;

- ICT has high potentials as a value-adding industry for Africa's industrialisation.

Potentials are huge, but, as always, the first step is the most difficult one – how to achieve a clear understanding and conviction among the country top management that the digital transformation is the right thing to initiate in your country?

This book is the answer why and how a country must start its digital transformation. This book discusses all the aspects of the public services digital transformation with the detailed case of Nigeria.

- Vision, goals, theme and guiding principle of digital transformation
- Public services digital transformation
- Success factors in digital transformation
- The future and shape of digital transformation
- Digital transformation approach
- Digital government trends

This book actually provides a methodological background which can be used by any country to justify and jump start its own national digital transformation.

Alexander Samarin wrote his first software program in the year 1973. He obtained a PhD (in computer graphics) in the year 1986. He worked for a variety of international clients in Switzerland, UK, France, Australia and Africa. He specialises in architecture, implementation and evolution of enterprise-wide solutions with the holistic use of enterprise architecture, business architecture, BPM, SOA, ECM, IT governance and IT strategy. In October 2009 he published a book "Improving Enterprise Business Process Management Systems". Since August 2013 he works as a consulting enterprise architect for implementation of digital uber-complex socio-technical systems.

References

[1] Driek D, Ewen Duncan et al, Six building blocks for creating a high performance digital enterprise, http://www.mckinsey.com/business-functions/organization/our-insights/six-building-blocks-for-creating-a-high-performing-digital-enterprise, accessed April 2015

[2] Creative Destruction Whips through Corporate America, http://www.innosight.com/innovation-resources/strategy-innovation/upload/creative-destruction-whips-through-corporate-america_final2015.pdf accessed April 2015.

[3] Schwab Klaus, The Fourth Industrial Revolution, World Economic Forum, 2016, page 2-3.

[4] David Chinn et al World-Class government – Transforming the UK public sector in an era of austerity: Five lessons from around the world   Discussion paper, March 2016.

[5] UN E-government Survey report 2014, https://publicadministration.un.org/egovkb/en-us/Reports/UN-E-Government-Survey-2014

[6] Delivering public services in a connected world, How Cisco can help shape your digital strategy, https://www.cisco.com/assets/global/UK/public_sector/new/pdfs/DigitalPublicServicesWebSinglePageV1.pdf

[7] World Economic Forum pushing for digital transformation in Africa, https://asokoinsight.com/news/world-economic-forum-pushing-for-digital-transformation-in-africa

[8] Marc Saxer, Shaping The Great Digital Transformation, https://www.socialeurope.eu/2015/05/shaping-great-digital-transformation/

[9] Nigeria Population Review, http://worldpopulationreview.com/countries/nigeria-population/

[10] Nations economy compared, http://www.nationmaster.com/country-info/stats/Economy/Economy/Overview accessed April 2015

[11] World Bank Data Projection, http://data.worldbank.org/country/nigeria

[12] ECOWAS, http://www.ecowas.int

[13] GOPA International (2015) The Nigerian Energy Sector: An Overview with a Special Emphasis on Renewable Energy, Energy Efficiency and Rural Electrification, June 2015, available at https://www.giz.de/en/downloads/giz2015-en-nigerian-energy-sector.pdf

[14] World Bank 2015 GDP (World Bank Estimate), Debt Management Office (DMO)

[15] Digital In Nigeria infographics, http://wearesocial.com/sg?s=Nigeria

[16] Internet Live Stats, Elaboration of data by *International Telecommunication Union (ITU), World Bank, and United Nations Population Division, http://*www.InternetLiveStats.com

[17] Olaopa T (2014) The Centenary of the Nigerian Civil Service, available at http://www.premiumtimesng.com/opinion/154385-centenary-nigerian-civil-service-tunji-olaopa.html

[18] Kayode J et al (2013) Corruption and service delivery: the case of Nigerian public service, available at http://www.wudpecker-researchjournals.org/WJPA/pdf/2013/July/Kayode%20 et%20al.pdf

[19] Helen Chapin Metz, ed. Nigeria: A Country Study. Washington: GPO for the Library of Congress, 1991. Available at http://countrystudies.us/nigeria

[20] N165 billion monthly salaries of federal civil servants not sustainable, http://www.premiumtimesng.com/news/top-news/202961-n165-billion-monthly-salaries-federal-civil-servants-not-sustainable-adeosun.html   accessed May 06, 2016

[21] Faseluka Olugbenga Peter 2015, Civil service administration and effective service delivery for development, available at http://guardian.ng/features/civil-service-administration-and-effective-service-delivery-for-development/

[22] Sustainable development and the imperatives of public service reform in Nigeria, Bureau of Public Service Reform (2009)

[23] Ayawei P, Abila S, A T Kalama (2015) Corruption and corrupt practices in Nigeria: an agenda for taming the monster, available at https://independent.academia.edu/ProsperAyawei/Activity

[24] World Development Report, Making Services Work for Poor People (WDR 2004)

[25] Okojie O (2009) Decentralization and public service delivery in Nigeria available at http://www.ifpri.org/publication/decentralization-and-public-service-delivery-nigeria

[26] Rasul I, Dogger G (2015), Management of Bureaucrats and Public Service Delivery: Evidence from the Nigerian Civil Service, available at http://www.ucl.ac.uk/~uctpimr/research/CSS.pdf

[27] Trading Economics http://www.tradingeconomics.com

[28] Alter R et al (2015) Delivering public services for the future, European center for government transformation

[29] How I became running mate to Buhari, http://www.vanguardngr.com/2016/05/became-running-mate-buhari-osinbajo/

[30] Bhaskar Chakravorti, Christopher Tunard, Rav Shankar Chaturvedi, Digital Plant: readying for the rise of the e-consumer, Institute of Business in the global context, The Fletcher School, Tufts University http://fletcher.tufts.edu/~/media/Fletcher/Microsites/Planet%20eBiz/Digital%20Planet%20-%20Executive%20Summary.pdf

[31] The Kenya National ICT Master Plan https://www.kenet.or.ke/sites/default/files/Final%20ICT%20Masterplan%20Apr%202014.pdf accessed March 10, 2016

[32] Civil Registration – Digitization Documentary 2013, https://www.youtube.com/watch?v=XvkPDUNGzEU

[33] ICT Authority, Kenya Transport Management system, http://www.icta.go.ke/transport-integrated-management-systems-tims/ and http://www.icta.go.ke/new-technology-modernize-road-transport-sector/

[34] Andrew Waittu, Digital Transformation critical to advance government transparency and compliance across Africa, http://www.cio.co.ke/blog/digital-transformation-critical-to-advance-government-transparency-and-compliance-across-africa

[35] Christi, Susanne; Barberis, Janos (2016). The FINTECH Book: The Financial Technology handbook for Investors, Entrepreneurs and Visionaries

[36] Innovative Digital Lessons from Kenya, http://www.dw.com/en/innovative-digital-lessons-from-kenya/a-18507104

[37] Winfred Kuria, Kiambu County emerges the leading county in the Kenyan Government's digital transformation, 2015, http://www.kachwanya.com/2015/08/03/kiambu-county-emerges-the-leading-county-in-the-kenyan-governments-digital-transformation/

[38] Use of IT within Austrian Justice, https://www.justiz.gv.at/web2013/file/8ab4ac8322985dd501229ce3fb1900b4.de.0/itbrosch%C3%BCre-en.pdf

[39] Max Goldbart, How did you…Seamlessly digitize land and property searches? http://www.civilserviceworld.com/articles/interview/how-did-you%E2%80%A6-seamlessly-digitise-land-and-property-searches

[40] Jacobs Edo, Operation Digital Evolution Transformation, 2015. https://www.academia.edu/14358119/Operations_Digital_Evolution

[41] Sofia Lotto Persio, Blockchain: paving the way for innovation in trade finance, http://www.gtreview.com/news/global/magazine-blockchain-paving-the-way-for-innovation-in-trade-finance/

[42] A digital government perspective, Accenture Technology Vision 2015 Delivering Public Service for the Future, techvision2015, https://www.accenture.com/us-en/~/media/Accenture/next-gen/public-service-technology-vision-trends-2015/downloads/Accenture-Public-Service-A-digital-government-perspective-US-letter.pdf

[43] Building a digital Government, PwC 2016 Digital Government Trends, www.pwc.com/ca/digitalgovernmenttrrends

[44] GSMA Mobile Economy 2016 Report, http://gsmamobileeconomy.com/

[45] Result 10 Blueprint: A strategy for digital public services, Internal Affairs 2014, New Zealand Government

[46] Mechthild Rohen, A Vision for Public Services, European Commission, 2013. Directorate-General for Communications Networks, Content and Technology, http://ec.europa.eu/information_society/newsroom/cf/dae/document.cfm?action=display&doc_id=3179

[47] Simone Di castri, Lara Gidvani, Enabling mobile money policies in Tanzania, A "test and learn" approach to enabling market-led digital financial services, 2014, http://www.gsma.com/mobilefordevelopment/wp-content/uploads/2014/03/Tanzania-Enabling-Mobile-Money-Policies.pdf

[48] Transform to a digital enterprise, Hewlett-Packard Enterprise http://h10131.www1.hpe.com/campaign/gov-transformation/

[49] Satoshi Nakamoto, Bitcoin: A peer-to-peer Electronic Cash System. This paper detailed methods of using a peer-to-peer network to generate what was described as "a system for

electronic transactions without trust, https://bitcoin.org/bitcoin.pdf

[50] Melanie Swam, Blockchain: Blueprint for a new economy, O'Reilly 2015

[51] Bates, L. (2016, January 16). Bitland Global White Paper. Retrieved from http://bitland.world

[52] Blockchain Technology: Possibilities for the US Postal Service Report Number RARC-WP-16-011 https://www.uspsoig.gov/sites/default/files/document-library-files/2016/RARC-WP-16-001.pdf

[53] Bitcoin and Blockchain Revolution for the price of one https://gendal.me/2015/07/23/bitcoin-and-blockchain-two-revolutions-for-the-price-of-one/

[54] Grace Caffyn, Everledger Brings Blockchain Technology to fight against diamond theft, http://www.coindesk.com/everledger-blockchain-tech-fight-diamond-theft/

[55] Digital Transformation, http://4-advice.net/digital-transformation/

[56] Common shared services candidates, http://www.imanet.org/docs/default-source/thought_leadership/operations-process-management-innovation/implementing_shared_service_centers.pdf?sfvrsn=2

[57] Mel Poluck, Common procurement solution for London schools saves £300 million, http://www.ukauthority.com/UKA-Local-Digital/entry/6260/common-procurement-solution-for-london-schools-saves-%C2%A3300-million

[58] Draft National Strategy for Public Sector Reforms, Nigerian Government available at http://bpsr.gov.ng/index.php/publications/bpsr-resources/strategy-documents

[59] Donella, Meadows, Thinking in Systems: A Primer, Chelsea Green Publishing, 2008 (page 17)

[60] Capgemini roadmap for a billion dollar organizations, https://www.capgemini.com/resource-file-access/resource/pdf/Digital_Transformation__A_Road-Map_for_Billion-Dollar_Organizations.pdf

[61] Scotland's Digital Services, http://www.gov.scot/Topics/Economy/digital/digitalservices

[62] Emma Weston, New National programme to deliver basic digital skills, 2016 https://doteveryone.org.uk/blog/2016/02/new-national-programme-to-deliver-basic-digital-sk/

[63] Nigerian National Policy for Information Technology (IT) "USE IT" http://www.researchictafrica.net/countries/nigeria/Nigerian_National_Policy_for_Information_Technology_2000.pdf

[64] Nchuchuwe F and David O A (2015) Challenges and Prospects of Implementing E-governance in Nigeria

[65] NeGST, www.negst,com.ng/index/php/about-us

[66] Fraser-Moleketi, G and Senghor, D 2011 E-governance and Citizen Participation in West Africa: Challenges and Opportunities. New York: UNDP in collaboration with PIWA-West Africa

[67] United Nations Public Administration Country Studies available at https://publicadministration.un.org/egovkb/en-us/Data/Country-Information/id/125-Nigeria

[68] Public Sector digitisation; the trillion-dollar challenge, available at http://www.mckinsey.com/business-functions/business-technology/our-insights/public-sector-digitization-the-trillion-dollar-challenge

[69] Emeka Aginam, Emma Elebake, Way forward for telecoms sector growth, by NCC operators, http://www.vanguardngr.com/2012/04/way-forward-for-telecoms-sector-growth-by-ncc-operators/

[70] Cem Dilmegani, Bengi Korkmaz, and Martin Lundqvist, Public-sector digitization: The trillion-dollar challenge, December 2014, http://www.mckinsey.com/business-functions/business-technology/our-insights/public-sector-digitization-the-trillion-dollar-challenge

[71] Emmanuel Adetula, The Naked Truth, Xlibris Corporation, 2011, page 40.

[72] Digitization of government brings wide benefits, available at http://www.itp.net/601968-digitization-of-govt-services-brings-wide-benefits-says-accenture

[73] Deloitte digital Democracy Survey, 9th edition https://www2.deloitte.com/content/dam/Deloitte/global/Documents/Technology-Media-Telecommunications/gx-tmt-deloitte-democracy-survey.pdf

[74] Bank 2020 Blockchain powering the internet of value, EVRY whitepaper, https://www.evry.com/globalassets/insight/bank2020/bank-2020---blockchain-powering-the-internet-of-value---whitepaper.pdf

[75] Anazodo Rosemary et al Civil Service Reforms in Nigeria: The journey so far in service Delivery American Journal of Social and Management Sciences, page 17-29, 2012

[76] Edo Jacobs (2015) Nigerian Civil Service Digital Transformation available at http://www.jacobsedo.com/1622-2/

[77]] Ahonen and O'Reilly 2007; ITU 2005; Kelly, Gray, and Minges 2003; Korea National Statistics Office 2007.

[78] John Barrdear & Michael Kumhof, The macroeconomics of central bank issued digital currencies, July 2016, http://www.bankofengland.co.uk/research/Documents/working-papers/2016/swp605.pdf

[79] Strategies for a peak body for an Australian national electronic authentication framework [including a Public Key Authentication Framework (PKAF)], a report prepared for the national office for the information economy, 1997. https://www.finance.gov.au/agimo-archive/__data/assets/file/0020/21629/npkiworkingpartyreport.pdf

[80] Welcome to Central Public Procurement Portal, https://eprocure.gov.in/cppp/

[81] Eleanor Whitehead & Michael Martins, Building a digital Nigeria, An Economic Intelligence Unit report commissioned by Accenture, 2016.

[82] Ross Alec, The Industries of the Future, Published by Simon & Schuster, (2016)

[83] Alan Morrison, Blockchain and smart contracts automation: How smart contracts automate digital business, PwC *Technology Forecast* series. https://www.pwc.com/us/en/technology-forecast/blockchain/digital-business.html

[89] Oronsaye Committee report as submitted to the President of Nigeria in Abuja. http://agora.nigeriagovernance.org/wp-content/uploads/2015/04/Oronsaye-Main-Report.pdf

[90] The World Factbook: 2010 Edition (CIA's 2009 Edition

[91] http://www.iedp.com/articles/prepare-for-digital-business-transformation

[92] Achieve Digital Excellence https://econsultancy.com

[93] Gerald C. Kane, Doug Palmer, et al, Strategy, not Technology, Droves Digital Transformation, MIT SLOAN Management Review, 2015 http://sloanreview.mit.edu/projects/strategy-drives-digital-transformation/

[94] Kathy O' Connell, Kelvin Delaney and Robert Moriarty, Digital Transformation: Disrupt to Win, Cisco, June 2015. http://www.connectedfuturesmag.com/Research_Analysis/docs/digital-business-transformation.pdf

[95] Joseph Bradley, Lauren Buckalew, Jeff Loucks and James Macaulay, Internet of everything in the Public Sector: Generating Value in era of change – top 10 insights, Cisco, 2013 http://internetofeverything.cisco.com/sites/default/files/pdfs/public_sector_jurisdiction_top_ten_insights_final.pdf

[96] MIT SMR and Deloitte, "Strategy, Not Technology, Drives Digital Transformation", July 2015. Survey of more than 1,200 government officials from over 70 countries on digital transformation and interview of another 140 government leaders and outside experts by Deloitte digital business and MIT SMR

[97] GDS UK, "How digital and technology transformation saved £1.7bn last year", https://gds.blog.gov.uk/2015/10/23/how-digital-and-technology-transformation-saved-1-7bn-last-year/, October 2015

[98] McKinsey & Co., "Public-sector digitization: The trillion-dollar challenge", December 2014

[99] EIU and Accenture, "Building a digital Nigeria", March 2016

[100] Deloitte, "The ascent of digital – Understanding and accelerating the sector's evolution", 2015

[101] Deloitte University Press, "The journey to government's digital transformation", October 2015

[102] BCG perspectives, "How to Jump-Start a Digital Transformation", September 2015

[103] Opengovernmentprtnership.org, "Republic of Korea national action plan on open government partnership", July 2015

[104] J. Langford and R. Roy, "Integrating service delivery across levels of government- cases studies of Canada and other countries," University of Quebec, 2008

[105] Technologytimes.ng, "Digitization of Nigerian public sector, a national emergency, CEO of Emerging Platform says", Nigeria International Technology Exhibition and Conference (NITEC) 2016 in Lagos, June 2016

[106] New Zealand Government ICT, www.ict.govt.nz/programmes-and-initiatives/information-privacy-and-security, "Information Privacy and Security", August 2015

[107] Whitehouse.gov, www.whitehouse.gov/sites/default/files/omb/egov/digital-government/digital-government.html, "Digital Government", August 2015

[108] The Independent, www.independent.co.uk/news/business/news/london-launch-for-d5-alliance-of-digital-nations-9909374.html, ""London launch for D5 alliance of digital nations", August 2015

[109] Christine Zhen-Wei Qiang and Carlo Rossotto, Economic Impacts of Broadband, Information and Communications for Development, page 40, 2009.

[110] Economic Commission for Africa, Partnership on Measuring ICT for Development, "Framework for a Set Of E Government Core Indicators", December 2011

[111] Economic Commission for Africa, "Framework for a set of e-government core indicators", March 2012

[112] Capgemini, IDC, Rand Europe, Sogeti and DTi, "Digitizing Public Services in Europe: Putting ambition into action", December 2010

[113] Gov.uk, GDS, "Performance Framework", https://www.gov.uk/design-principles/performanceframework, July 2012

[114] Enrique Quevedo, Daniel Besquin, Michelle Read, "Digital Metrics Playbook – Measuring your online branding strategies", 2015

[115] GDS Blog, "Beating performance anxiety", https://gds.blog.gov.uk/2013/03/28/beating-performance-anxiety/, March 2013

[116] Marcel Britsch, The Digital Business Analyst "Objectives, KPIs and Metrics - assessing project value", http://www.thedigitalbusinessanalyst.co.uk/2014/05/objectives-kpis-and-metrics-assessing.html, May 2014

[117] Digital 21 Strategy, The Government of the Hong Kong Special Administrative Region

[118] Ibrahim H. Osman, Abdel Latef Anouze, Bizhan Azad, Lina Daouk, Fouad Zablith, "The Elicitation of Key Perfor-

mance Indicators of E-government Providers: A Bottom-Up Approach", October 2013

[119] I-SCOOP, "Digital transformation and information management: enabling change", http://www.i-scoop.eu/digi-
tal-transformation/digital-transformation-and-informa-
tion-management-enabling-change/

[120] Global Centre for Digital Business Transformation, IMD and
Cisco, "Digital Business Transformation – A Conceptual
Framework", June 2015

[121] Gartner, "Digital Business KPIs: Defining and Measuring
Success", March 2016

[122] GDS UK, "GDS business plan April 2014 to March 2015", July
2014

[123] GDS UK, "Capabilities Plan for the Civil Service", April 2013

[124] Capgemini Consulting, "The Digital Talent Gap Developing
Skills for Today's Digital Organizations", 2013

[125] Transforming public services- the next phase of reform, www.
scotland.gov.uk, 2006

[126] Implementation of the virtual data embassy solution summary
report, Microsoft, 2015

[127] Recommendation of the Council on Digital Government Strategies, www.oecd.org, adopted by OECD Council on July 15,
2014.

Glossary

Acceptable Use Policy (AUP): An AUP is a set of rules that define the ways in which ICT facilities can and cannot be used in a business or educational institution, including a description of the possible sanctions that can be applied if a user breaks the rules. Two of the most important topics covered by an AUP are (i) **e-safety** and (ii) **awareness of and compliance with copyright**.

Access: The name of a Database program forming part of the Microsoft Office suite of programs.

Accessibility: The fundamental issue regarding *accessibility* is that everyone should have access to the services provided by ICT, e.g. computer programs, Email and the World Wide Web, regardless of any visual, auditory, or other physical impairment they might have. Assistive Technology may be employed to increase access to such services, e.g. Text To Speech (TTS) screen readers, screen magnifiers, speech recognition systems, hearing assistance devices, etc. Designers of computer programs and websites need to take account of accessibility when choosing colours, fonts and font sizes, etc:

Analogue: The basic meaning of analogue is "something that corresponds to something else". For example, in the context of equipment used for recording and playing back sound, analogue refers to the way in which the sound is recorded and reproduced. If you look closely at the groove of a 33 rpm vinyl gramophone record you will see that it is essentially a continuous wave, an undulating series of "hills". These "hills" correspond to the nature and volume of the sound that has been recorded. As the stylus of the record player moves along the wave it produces vibrations that are amplified and converted into sound. A parallel can be drawn with radio transmissions, where the sound signals are transmitted in the form of invisible waves. Early mobile phones worked in a similar way. Older tape recorders and videocassette recorders are based on the same principle, except that the signals representing the sound and moving images are imprinted onto a plastic tape coated with a magnetic powder. All analogue recordings suffer from background noise, and the quality of reproduction gradually degrades as the record or tape wears out. If the recording is copied, the copy will not be as good as the original, regardless of the quality of the equipment used to copy it.

API: Abbreviation for Application Programming Interface. API is a so-called protocol of communication that enables different computer programs to communicate with one another. A good API makes it easier to develop a program by providing all the building blocks that the programmer needs. Although APIs are designed for programmers, they are also good for program users insofar as they guarantee that all programs using a common API will have similar interfaces. This makes it easier for users to learn new programs.

App: Abbreviation for Application.

Application: A computer program or a suite of computer programs that performs a particular function for the user, such as a word-processor, e.g. Microsoft Word, or a range of functions, such as Microsoft Windows or Microsoft Office. Commonly abbreviated to app, especially in the context of Web 2.0 and Mobile Assisted Learning (MALL) apps.

Archive: Used to describe documents or files that are not immediately needed but which should not be completely discarded.

An archive may be stored on an external Hard Disc, CD-ROM, DVD or other Storage Device. Also used to describe stored messages that have been contributed to discussion lists or blogs. Also used as a verb.

Artificial Intelligence (AI): The ability of a computer to mimic human attributes in finding a solution to a problem.

Bit: Contraction of binary digit. A bit is the smallest measurement unit of computer memory or data transmission speed, e.g. via a Modem.

Blog: Contraction of the term Weblog. A blog is essentially a website that contains discrete pieces of information posted by different users. New items of information are usually entered by contributors via a simple form, following the introduction of each new theme by a person who initiates the blog, and then submitted to the site, where they may be filtered by an administrator before being posted. A blog can contain news items, short essays, annotated links, documents, graphics, and multimedia. These posts are usually in reverse chrono-logical order and often take the form of a journal or diary. A blog is normally accessible to any Internet user, but closed blogs may also be created, e.g. to document the thoughts and experiences of a group of students or to provide a means of communication between teachers and students following a particular course.

Blogger: Normally used to refer to someone who blogs, i.e. who regularly writes blogs. Also used to describe a service that provides Web-based tools used by individuals to create a Blog or Weblog.

Broadband: A general term used to describe a high-speed connection to the Internet. Connection speed is usually measured in Kbps (kilo-bits per second) and Mbps (megabits per second). Typically, a home user will have a broadband connection using an ADSL telephone line running at 2Mbps to 8Mbps. Educational institutions ideally need a symmetric connection of at least 8Mbps to ensure smooth trouble-free connections to the Internet when large numbers of students are accessing the Internet all at once.

Browser: A software package installed on the hard disc of your computer that enables you to access and to navigate the World Wide Web - to "surf the Web" in colloquial terms.

Byte: A measurement of computer memory or disc capacity. A byte comprises 8 *bits.*

Comms: Short for *communications*, as in Information and Communications Technology (ICT). Used to refer to ways in which computer systems communicate with one another, e.g. via a cable, a telephone line, satellite or wireless.

Compatiblity: Pieces of hardware and/or software which are capable of being used together are described as *compatible.*

Computer Program: A set of instructions that the computer carries out in sequence to perform a given task. Programs are written in English-like programming languages (e.g. C, Pascal), and are then converted into binary machine instructions via a compiler or an interpreter.

CPD: Abbreviation for Continuing Professional Development.

CPU: Abbreviation for Central Processing Unit.

Crash: A term describing what happens to hardware or software when it suddenly fails to work properly. The commonest symptom of a crash is the "frozen screen", i.e. when the keyboard and/or mouse goes dead with the result that nothing can be typed and the Cursor cannot be moved around the screen. Modern computers typically crash several times a day. Most crashes are not serious and are simply the result of faulty programming - i.e. most programming. Some kinds of crashes can be symptomatic of more serious problems, however, and should be investigated if they keep occurring. Operating systems themselves, e.g. Microsoft Windows, are particularly prone to crashes. See Operating System, Windows.

Cyberspace: William Gibson coined this phrase in his novel *Neuromancer*, first published in 1984 - some years before the World Wide Web was invented: "Cyberspace. A consensual hallucination experienced daily by billions of legitimate operators, in every nation, by children being taught mathematical concepts... A graphic representation of data abstracted from the banks of every computer in the human system. Unthinkable complexity. Lines of light ranged in the nonspace of the mind, clusters and constellations of data. Like city lights, receding..." Today the word *cyberspace* is used to refer to the world of the Internet, more specifically the World Wide Web.

Cybersquatter: A term normally used to describe someone who registers the name of a popular Web address - usually a company name - with the intent of selling it to its rightful owner at a high price. Cybersquatters also watch out for registered domain names that become available when the owner has no further use for them, goes bankrupt, or simply forgets to pay their registration renewal fees. This can lead to perfectly harmless and legitimate sites being transmogrified overnight into sites containing offensive material.

Data: Strictly speaking the plural of "datum", but now usually considered as a collective noun in the singular, with the plural form "data items" or "items of data". Data is information in a form which can be processed by a computer. It is usually distinguished from a *computer program*, which is a set of instructions that a computer carries out. Data can be text or sets of figures on which a computer program operates.

Database: A structured collection of data that can be used for a variety of purposes. Databases are usually stored on a Hard Disc inside your computer, on a CD-ROM, or at a website. A database may contain data relating to staff employed by a company or to students at an educational institution. Databases can also contain bibliographies, glossaries, vocab lists, etc. In order to set up and manage a database you need a database program such as Microsoft Access.

Desktop: The main workspace in Windows is often referred to as a desktop, which is displayed on the screen that you see when Windows is started. This electronic desktop is a metaphor for the top of a real desktop, where jobs to be done are laid out in different folders symbolised by Icons, i.e. small images. Users open and work with programs by clicking on the icons on the desktop, and they can also store shortcuts to documents or websites there.

Desktop Computer: A desktop computer is one that is designed to sit permanently on a desk, as opposed to a portable computer, e.g. Laptop Computer, Netbook, Notebook Computer and Tablet Computer, all of which can easily be carried around.

Digital: The essential meaning of this term is "based on numbers". The modern computer is a typical example of digital technology, so are CD-ROMs, DVD-ROMs, audio CDs and video DVDs, on which numbers are coded as a string of tiny pits pressed into a plastic disc. When a CD audio recording or a DVD video recording is played back, using equipment incorporating a laser as a reading device, the exact numeric values are retrieved and converted into sound or images. Digital recording is relatively free from noise and interference and gives a very high quality of reproduction. Data (including audio and video) or programs stored on CD-ROM or DVD can be read by a computer in a similar way. There are two major benefits to digital technology. Firstly, digital technology - because it is based on numbers - is more precise. Secondly, digital technology is becoming cheaper and more powerful. Digital technology is now used in radio and TV broadcasts. Digital recordings made from any source (audio- or videocassettes, television, radio, Internet, satellite TV, microphone or Camcorder) can be edited easily, then stored on a computer's Hard Disc, CD-ROM, DVD, Flash Drive, Memory Stick, etc. They can be copied without quality loss and, more significantly, can be used by more than one learner at the same time. See the contrasting term Analogue.

Digitise / Digitize: To translate into a digital form, i.e. numbers. For example, scanners *digitise* images by translating them into *bitmaps*, i.e. thousands of individual dots or *pixels*. It is also possible to digitise sound and video by Sampling at discrete intervals. To digitise sound, for example, a device measures a sound wave's characteristics many times per second and converts them into numeric values which can then be recorded.

E-learning: *E-learning* (electronic learning) has become a buzz-word in recent years, but it is widely misunderstood and often associated with a limited view of e-learning. Ask a dozen people what they understand by e-learning and most will probably say that it involves using a computer to access materials on the Web or to follow a distance-learning course using a Virtual Learning Environment (VLE). Here is the definition given in the UK government's consultation document Towards a unified e-learning strategy, July 2003:

If someone is learning in a way that uses Information and Communications Technologies (ICTs), they are using e-learning. They could be a pre-school child playing an interactive game; they could be a group of pupils collaborating on a history project with pupils in another country via the Internet; they could be geography students watching an animated diagram of a volcanic eruption their lecturer has just downloaded; they could be a nurse taking her driving theory test online with a reading aid to help her dyslexia - it all counts as e-learning.

Electronic Mail: See Email.

Electronic Whiteboard: More commonly referred to as an Interactive Whiteboard these days.

Email: Contraction of Electronic Mail. A system for creating, sending and receiving messages via the Internet. In order to send and receive email messages you have to register with an Internet Service Provider (ISP) that provides an email service and have email software such as Outlook or Eudora installed on your computer. Many ISPs also offer a Webmail facility, which provides an alternative means of creating, sending and receiving email messages using your Web Browser.

Encryption: A system of coding that helps prevent access to private information on computer networks or on the Web.

End-user: The final user of a piece of Software or Hardware, i.e. the individual person for whom the product is created, as distinct from the people who create and produce the product.

File: A file in computer jargon can be used to describe many different things. It may be a Computer Program, a document file created with a Word-processor, an image file, an audio file, a video file, etc. Think of it in the same way as you would think of a file in a filing cabinet. A file has a name that describes what it is, and the file is stored in a place where you can easily find it. Files are usually grouped together on a computer's Hard Disc in directories or folders and, as well as their names, they usually have a three-letter Extension that tell you what their function is or what they contain, e.g. fwtt. exe is a program, mystory.doc is a Word document, sally.jpg is a picture, and mydog.mpg is a video file. Files may also be stored on CD-ROMs, DVDs andFlash Drives.

Filename: The name of a File on a computer.

File Permissions: Files stored on a computer usually have *permissions* governing which users are allowed to read, amend or execute them. This is particularly important in a a school, college or university network environment, where teachers and lecturers may have the permission to amend certain files, e.g. documents that they have created, but students are only allowed to read them. File permissions are usually determined by network managers.

File Transfer Protocol: See FTP.

Firewall: A *firewall* is a software package that sits between your computer and your Internet connection, keeping an eye on the traffic going to and fro. If anything suspicious appears, such as an unauthorised attempt from a remote computer to write information to your hard disc or to send information from your computer to a remote computer, it will block it and warn you. Firewalls have become essential these days because of the frequent attempts being made by *hackers* to grab confidential information from computers all around the world, e.g. your bank or credit card details, which may be stored in a file somewhere on your computer. Any computer is vulnerable while it is connected to the Internet. The author of this paragraph writes from personal experience: two attempts have been made by hackers to grab passwords from his computer. Both attempts were fortunately spotted by his Internet Service Provider and blocked, so no damage was done. If you access the Internet via a computer in a public or commercial organisation your ICT services department has almost certainly installed a firewall, but if you access the Internet via your personal computer then you should make sure that you install your own firewall. In addition you should install an *anti-virus package.*

Folder: An alternative word for a *directory* and which has become more common since the introduction of Windows. It describes a location on a disc which contains a set of related files. A folder can be divided into sub-folders.

Font: The terms font (also spelt fount) and typeface are often confused or interchanged. Font refers to a complete collection of letters, numerals, symbols and punctuation marks that have common characteristics, including their style and size. The two commonest fonts are Times New Roman, a Serif font, which is characterised by cross-lines that finish off the stroke of each letter, and Arial, a Sans Serif font that has no cross-lines. Typeface is the name given to the style of a particular set of letters, numerals, symbols and punctuation marks.

Formatting: The process of preparing a writeable disc for use. Formatting creates a structure on the disc which enables it to hold data.

GHz: Abbreviation for GigaHertz.

GIF: Abbreviation for Graphic Interchange Format. A file format used for storing simple graphics. GIF files use a palette of 256 colours, which makes them practical for almost all graphics except photographs. Generally, GIF files should be used for logos, line drawings, icons, etc, i.e. images that don't contain a rich range of colours. A GIF file containing a small number of colours tends to be quite small, but it will be big if the image has a wide range of colours, e.g. a photograph.GIF files are commonly used for storing images on the Web. GIF files are also suitable for storing animated (i.e. moving) images.

Gigabyte: Usually abbreviated to *GB*, or *gig* in common computer parlance. A unit of measurement of computer memory or disc capacity = 1,073,741,824 bytes.

GigaHertz: Usually abbreviated to *GHz*. A unit of measurement relating to the Clock Speed of a computer or, put simply, a measurement of how fast its Central Processing Unit (CPU) runs. Typical clock speeds of modern computers range from 500MegaHertz (500MHz) upwards. Faster clock speeds are normally expressed in GigaHertz (= 1000MHz)

Graphical User Interface (GUI): An Interface, i.e. a software package, that enables human beings to control what happens on their computers. A GUI consists of graphical elements known as *icons* and enables the user to run programs and to carry out other operations such as copying information from one Folder to another, deleting files, etc by clicking on these icons, opening and shutting windows and dragging and dropping with a mouse. Microsoft Windows and the much older Apple Mac interface are GUIs. Contrasted with Character User Interface (CUI), an older type of interface which required the user to control the computer by typing commands at the Keyboard.

Hacker: A person who spends their time trying to gain access to information stored on other people's computers all around the world. Some hackers are just harmless browsing types, but other have more invidious aims such as grabbing details of your credit cards or bank account, which may be stored in a file somewhere on your computer. If you access the Internet regularly you should consider installing a Firewall to protect yourself against hackers.

Hardcopy or **Hard Copy:** Printed output from a computer, as opposed to output on screen.

Hard Disc: A *hard disc* consists of a single rigid magnetic disc or a set of such discs enclosed within a metal case, i.e. a *hard disc drive*, which is mounted internally in your computer and is used for storing the computer programs and data that it needs in order to work. External hard disc drives can also be obtained for additional storage capacity or backup storage. Hard discs can contain vast amounts of data, usually measured in *gigabytes*.

Hardware: The physical elements of a computer system - the bits you can see, touch, drop, kick or fall over. Contrasted with Software.

HDD: Abbreviation for Hard Disc Drive. See Hard Disc.

HTML: Abbreviation for Hypertext Markup Language. The coding system used for creating pages on the World Wide Web. HTML enables the author to control how the page appears and to insert Hypertext links within one Web page or to other pages anywhere on the Web. Nowadays most Web authors and designers use an Authoring Tool such as Front Page or Dreamweaver to create World Wide Web pages. Web page files end with the Extension .htm. or .html.

HTTP: Abbreviation for Hypertext Transfer Protocol. The transfer method (*protocol*) used by the World Wide Web to transmit and receive Web pages.

Hyperlink: A contraction of hypertext link, the essence of Hypertext and the HTML language used for creating pages on the World Wide Web. In a Web document a hyperlink can be a sequence of letters or an image. By clicking on the area designated as a *hyperlink* by the person who created the Web page, it is possible to jump quickly to another part of the page, a different page on the same website, or to a completely different website.

Hypermedia: The extension of the *hypertext* concept to *multimedia*, describing the combination of multimedia information (text, images, audio, video, etc) in a meaningful configuration, which is especially useful for teaching and learning.

Hypertext: A system for the non-sequential presentation of text, the fundamental concept of the World Wide Web, whereby the user can jump from one part of a text to another, from one Web page to another, or from one website to another, by clicking on highlighted (and usually underlined) *hyperlinks*. The concept of *hypertext* predates the Web by many years. Vannevar Bush is credited with inventing the concept of *hypertext* in his article "As we may think", which was written as early as 1945 and describes an imaginary machine called "Memex" - essentially a hypertext device that takes account of the way the human mind associates ideas and follows a variety of different paths rather than moving on sequentially

ICT: Abbreviation for Information and Communications Technology. ICT is the term that is currently favoured by most businesses and educational institutions. The "C" reflects the important role that computers now play in *communications*, e.g. by email, the Web, by satellite and cellphone (mobile phone). We always insist on the "s" at the end of *communications*, which is a term that predates computer technology and was originally associated with morse code, radio, etc and often abbreviated to *comms*.

Interface: An *interface* in computer jargon is a connection between two systems. It can be Hardware or Software. It may take the form of a plug, cable or socket, or all three, for example where a Printer or Scanner is connected to a computer, and then it's a hardware interface. There are also software interfaces that enable one program to link with another, passing across data and variables. The term interface, also known as user interface, also describes the software that is used to enable human beings to communicate with a computer, for example Microsoft Windows, which is a Graphical User Interface (GUI) in common use on personal computers.

Internet: The *Internet*, or simply "the Net", is a computer network connecting millions of computers all over the world. It provides communications to governments, businesses, universities, schools and homes. Any modern computer can be connected to the Internet using existing communications systems. Schools and universities normally access the Internet via their own educational networks, but private individuals usually have to take out a subscription with an Internet Service Provider (ISP). Although the Internet is in fact a network of networks, it appears to users as a network of individual computers. The Internet dates back to the group of interconnected networks that evolved from the ARPANET of the late 60's and early 70's. It has grown from a handful of interconnected networks into a huge network of millions of computers. The main Internet services of interest to language teachers are Email and World Wide Web.

Internet Explorer: A Browser produced by the Microsoft Corporation and supplied together with the Windows operating system.

Internet Service Provider (ISP): A company that provides a subscription service to enable you to access the Internet. An ISP has a network of computers permanently linked to the Internet. When you take out a subscription with an ISP they link your computer to their network, usually via an existing telephone line, but dedicated lines are also provided by some ISPs. ISPs also give you an Email address and space on the World Wide Web for setting up your own website.

Interpreter: Software which converts the human-readable Source Code of a program which has been written in a high-level programming language such as BASIC, one statement at a time, into machine instructions as the application is run. Interpreted applications need to be distributed with runtime programs and function libraries.

Intranet: A private network inside a company or educational organisation and used over its LAN (Local Area Network). A sort of local Internet. Contrasted with Internet, which is publicly available.

ISDN: Abbreviation for Integrated Services Digital Network. A type of digital telephone service, used for transferring large chunks of data to and from the Internet without a Modem. Gradually falling out of use these days with the introduction of ADSL broadband services. ISDN lines normally operate at 128 Kbps, which is faster than a standard 56Kbps Dial-up Modem but slower than an ADSL connection, which runs at a speed of at least 1Mbps.

ISP: Abbreviation for Internet Service Provider.

IT: Abbreviation for Information Technology. Essentially, technology relating to information processing, i.e. computer technology, but see also ICT, C&IT, both of which describe the converging of information technology and communications technology. The term IT is rapidly being replaced by ICT in order to reflect the important role that information technology plays in communications by email, the Web, satellites and mobile phones.

Java: A programming language, invented by Sun Microsystems, that is specifically designed for writing programs that can be downloaded to your computer through the Internet and immediately executed. Using small Java programs, called *applets*, Web pages can include functions such as animations, interactive sequences, etc. You need to set up your browser to enable it to interpret and run the Java applets. Java is similar to a programming language known as C++ but it has been considerably simplified. Not to be confused with Javascript.

Javascript: *Javascript* is a script language, a system of programming codes that can be embedded into the HTML code of a Web page to add functionality, e.g. interactive sequences, questionnaires, etc. Although it shares many of the features and structures of the full Java language, Javascript is essentially quite different and was developed independently.

Kb: Abbreviation for Kilobit.

KB: Abbreviation for Kilobyte. The single letter K is also used.

Kbps: Abbreviation for *kilobits per second*. A unit of measurement of data transmission speed, e.g. via a Modem.

LAN: Abbreviation for Local Area Network. A Network of computers at one site that provides services to other computers connected to it. A LAN is usually limited to an immediate area, for example the floor of a building, a single building or a campus. The most important part of a LAN is the Server that delivers software to the computers (also known as workstations or clients) that are connected to it. The server is usually the most powerful computer in the network Users of computers connected to a LAN can access their own files remotely and exchange information with the server and other users connected to the network.

Linux: A Unix-type Operating System, similar to Windows and the Apple Mac operating system. Linux was originally created by Linus Torvalds with the assistance of developers around the world. The Source Code for Linux is freely available to everyone.

Local Area Network (LAN): See LAN.

MAN: Abbreviation for Metropolitan Area Network. A network of computers located at different sites within a large fixed area, such as a city.

Megabit: Usually abbreviated to *Mb*. 1,024 *kilobits* or 1,048,576 *bits*, a unit of measurement, usually relating to data transmission speed.

Megabyte: Usually abbreviated to *MB*. 1,024 *kilobytes* or 1,048,576 *bytes*. A unit of measurement of computer memory or disc capacity. Roughly 180,000 words of text - an average-sized novel. See entry on Measurement Units. See Bit, Byte, Kilobyte,Gigabyte.

MegaHertz: Usually abbreviated to *MHz*. A unit of measurement relating to the Clock Speed of a computer or, put simply, a measurement of how fast its Central Processing Unit (CPU) runs. Typical clock speeds of modern computers range from 500MHz upwards. Faster clock speeds are normally expressed in GigaHertz or GHz (= 1000MHz).

Memory: Most people use this term to refer to a computer's temporary internal main memory or RAM. Memory may also refer to ROM (Read Only Memory), which is permanent and part of a computer system as supplied by the manufacturer.

Memory Stick: A small electronic card, also known as a *memory card*, which is inserted into a Digital Camera or Camcorder for storing photographs or movie files that can then uploaded to a computer. This term is also used as an alternative to Flash Drive.

Navigation: This describes the process of finding your way, i.e. *navigating*, around a series of menus within a computer program or finding your way around the World Wide Web by means of a Browser.

Notebook Computer: A type of Laptop Computer, but lighter and thinner - and therefore easy to carry around. See Netbook, an even smaller and lighter computer.

Online Learning: The use of the Internet to follow a course that usually results in the award of a diploma or certificate. Closely associated with the concept of E-learning, which often - but not necessarily - implies some form of *online learning*, i.e. usingEmail and the World Wide Web. E-learning, i.e. electronic learning, is a broader term, embracing the use of ICT in general in teaching and learning as well as online learning.

Open and Integrated Learning System (OILS): A variant of Integrated Learning System. The word *Open* adds an extra dimension, indicating that the user can access the system freely and leave it at any time.

Open Source: Used to describe Software that is provided free of charge, along with the original Source Code used to create it so that anyone modify it to improve it and work in ways that reflect their own preferences. Moodle is a typical example of open source software.

Operating System (OS): A suite of programs that starts up when you switch on your computer and manages and runs all the other programs installed on the computer. *Windows* is the *operating system* developed and produced by the Microsoft Corporation.

PDF: An abbreviation for Portable Document Format. This is a file type created by Adobe that allows fully formatted, documents to be transmitted across the Internet and viewed on any computer that has Adobe *Acrobat Reader* software - a proprietary software viewing program available for free at the Adobe website: http://www.adobe.com/uk/. Businesses and educational institutions often use PDF-formatted files to display the original look of their brochures or for publishing a complete magazine in electronic format. Using the full Adobe *Acrobat* software package, it is possible to create a high-quality piece of artwork or a brochure which preserves the look of the original, complete with fonts, colours, images, and formatting. Documents in PDF format can be published on the Web without having to be converted into HTML. PDF files can be distributed via email, CD-ROMs and local area networks. They can also contain hyperlinks, QuickTime movies and sound clips.

RAM: An acronym for Random Access Memory, referring to the dynamic memory in the silicon chips in a computer. RAM chips are the memory chips used as the temporary working area for running and developing programs. Data in RAM can be read and written to (i.e. changed) in microseconds, as opposed to the much slower data access times for discs, but RAM's contents disappear the moment the computer is switched off. The more RAM a computer has, the more flexibility the user has. RAM used to be measured in *kilobytes (KB)* but now it is usually expressed in *megabytes (MB)* and even *gigabytes (GB)*. The amount of RAM a PC has could crudely be thought of as its "mental capacity".

Random Access Memory (RAM): See RAM.

RGB: Abbreviation for Red Green Blue. The name given to the Additive Colour system that is used to display colours on computer screens, where red, green and blue light of varying intensities is combined to produce millions of other colours.

Read Only Memory (ROM): See ROM.

Search Engine: A search facility provided at a number of sites on the *World Wide Web*. Search engines enable the user to search the whole of the Web for key words and phrases and to locate related websites. This is a useful facility for locating information. See Section 4, Module 1.5, headed *Search engines: How to find materials on the Web*.

Semantic Web: The *Semantic Web* is not a new type of Web, but rather an extension of the Web whereby data available in different locations on the Web is linked together in a way that allows the user to search the Web in a more sophisticated way, e.g. by requesting information in forms such as "Tell me where I can find information about 21st-century writers who live within 50 miles of my home town": http://www.w3.org/RDF/FAQ. Listen to Sir Tim Berners-Lee on the BBC Today programme, 9 July 2008, talking about the Semantic Web: http://news.bbc.co.uk/today/hi/today/newsid_7496000/7496976.stm

Smartphone: A *smartphone* is an advanced mobile phone that offers a wide range of applications. In addition to functioning as a mobile phone smartphones can be used as a media player, a camera, a GPS navigation device and a Web browser - and in many other ways. Apple's **iPhone** is a typical example of a smartphone, using a touchscreen for typing and to run applications.

Social Media: Term used to describe a variety of Web 2.0 applications that enable people to share images, audio recordings and video recordings via the Web and to initiate discussions about them. See JISC's **Web2practice** video on **Blip TV**:http://web2practice.jiscinvolve.org/social-media/

Social Networking: A term applied to a type of website where people can seek other people who share their interests, find out what's going on in their areas of interest, and share information one another.

Software: The opposite to Hardware. A generic term describing all kinds of computer programs, applications and operating systems. Software is not tangible, being a set of instructions written in a Programming Language comprising a set of instructions that the computer executes. See Application, Computer Program.

Sound Card or **Soundcard:** A *card*, i.e. an electronic circuit board, inside a computer that controls output to speakers or headphones and sound input from a Microphone or other source. A sound card is essential for multimedia applications. Also known as Audio Card.

Source Code: The human-readable form of a *computer program*, which is converted into binary computer instructions by a *compiler* or *interpreter*.

Spam: Unsolicited email advertisements, the Internet equivalent of junk mail.A *spammer* is someone who sends out spam. A spammer can email an advertisement to millions of email addresses, news-groups, and discussion lists at very little cost in terms of money or time. The term *spam* comes from a sketch in the *Monty Python's Flying Circus* TV series.

Tablet Computer: A *tablet computer* is compact portable computer that makes use of a Touchscreen instead of a keyboard for typing and running applications. Apple's **iPad** is a typical example of a tablet computer.

Tag: *Tagging* has become more common in recent years as a result of the widespread use of Social Media for sharing images, audio record-ings, video recordings, website references, etc. *Tags* are labels that briefly describe the what the media or references are all about and help other people find them quickly. Tags are also used in HTML, to define how the onscreen text is rendered by the browser: for example the tag **<a href="http://www.ict4lt.org">ICT4LT</a>** in HTML appears as ICT4LT, with the tag hidden to the person viewing the Web page.

Unix: An Operating System widely used on large computer systems in corporations and universities, on which many *Web servers* are hosted. A PC version of *Unix*, called *Linux*, is becoming increasingly popular as an alternative to *Windows*.

Upload: To transfer a copy of a computer program, a text file, an image file, a sound file or a video file from one computer to another computer. This term can also be used to describe the process of: (i) transferring a photograph from a digital camera to a computer, (ii) transferring a sound recording from a digital sound recorder to a computer, and (iii) transferring a video recording from a Camcorder or Digital Camera to a computer.

URL: Abbreviation for Uniform Resource Locator. Also known as a Web Address. A URL contains the location of a resource on the Internet. A URL specifies the address of the computer where the resource is located, which may be the homepage of a website, e.g. http://www.ict4lt.org, or a sub-page, e.g. http://www.ict4lt.org/en/en_mod2-1.htm. The **http://** prefix can usually be omitted from a URL when it is entered in a Browser.

USB: Abbreviation for Universal Serial Bus.

User-friendly: Mainly used to describe Software. Software that is easy to use and offers guidance if the user does silly things is described as user-friendly. This term may also be applied to certain types of Hardware.

User Interface: See Interface.

Virtual Learning Environment (VLE): A VLE is a Web-based package designed to help teachers create online courses, together with facilities for teacher-learner communication and peer-to-peer communication. VLEs can be used to deliver learning materials within an institution or within a local education authority. They may even address a wider constituency, and can even be used on a worldwide basis. VLEs have certain advantages in terms of ease of delivery and management of learning materials. They may, however, be restrictive in that the underlying pedagogy attempts to address a very wide range of subjects, and thus does not necessarily fit in with established practice in language learning and teaching. For this reason some critics argue in favour of a less restrictive Personal Learning Environment (PLE). The two most widely used VLEs in language teaching and learning are Blackboard and Moodle.

Virtual World: A type of online three-dimensional imaginary world or game in which participants and players adopt amazing characters or *avatars* and explore the world, engaging in chat or playing complex games.

Virus: If you surf the Web, use email or Storage Media sent to you by other people, you need to be protected against virus invasions. A virus is a nasty program devised by a clever programmer, usually with malicious intent. Viruses can be highly contagious, finding their way onto your computer's hard drive without your being aware of it and causing considerable damage to the software and data stored on it. Viruses can be contracted from files attached to email messages, e.g. Microsoft Word files, or direct from the Web. Be very wary of opening an email attachment of unknown origin, as this is the commonest way of spreading viruses. Software used to protect your computer against the invasion of computer viruses is known as *anti-virus software.*

VoIP: Abbreviation for Voice over Internet Protocol, i.e. audio communication using the Internet instead of telephones. Skype and Ventrilo are examples of VoIP.

WAN: Abbreviation for Wide Area Network. A network of computers located at geographically separate sites. See LAN, MAN.

WAP: Abbreviation for Wireless Application Protocol. A system that enables you to browse online services, e.g. relating to information about the weather, traffic conditions, shopping, etc. via a special type of mobile phone. WAP is the mobile phone equivalent of the World Wide Web. Newer mobile phones include WAP browser software to allow users access to WAP sites.

Web 2.0: Contrary to what many people think, Web 2.0 is not a new version of the World Wide Web. The term arose as the name of a series of conferences, the first of which was held in 2004: http://www.web2summit.com. Essentially, Web 2.0 is an attempt to redefine what the Web is all about and how it is used, for example new Web-Based communities using Blogs, Podcasts, Wikis and Social Networking websites that promote collaboration and sharing between users - in other words, a more democratic approach to the use of the Web. In order to achieve this, Web-based applications have to work more like applications on your computer's hard disc, allowing you to use the Web in much the same way as you would use applications such as *Word* or *PowerPoint*. To what extent the concept of Web 2.0 is truly innovative is a matter of debate, as it is broadly in line with the concept of the Web as defined by its inventor, Tim Berners-Lee, way back in 1998.

Web Address: See URL.

Additional Reading Resource

THE CORRUPTION RIDDLE

BY ARNOLD OBOMANU

In spite of wide public condemnation of the corruption situation in our country, things seem to worsen. And contrary to the disclaimers a lot of us put out in public, we frequently find ourselves complicit in corrupt acts. Some may want to deny this but a friend puts it this way; he says someone who lives in Nigeria can get away with claiming not to have taken any bribe but he cannot successfully claim not to have given one! And this is not to glorify corruption but to accurately set the scene for taking a deeper look at the problem we are facing as individuals and as a nation.

Corruption is so ubiquitous in our society now that in a lot of situations, it has become the default way to get things done. And although the display of corruption and its attendant impunity grows on a daily basis we do not seem to know how to put it to rest. A lot of times when we decry corruption it is usually because we experience its negative consequences. And as corruption spreads, its impact is spreading across the strata of society even to the elite. Our elite may be able to fly out for treatment abroad but they must use the same chaotic airport as the rest of us. Even when they fly from city to city they have to at some point drive through the same cratered roads as the poorest Nigerian. And this provides a basis for everyone to position themselves on the same side of the table and against the corruption monster.

I find it interesting that Senator Maccido who died in the 2006 air crash was in the Senate when the 2005 Sosoliso crash was investigated. Or consider the fact that General Adisa, a former Minister of Works, who oversaw for some years the ministry responsible for road works across the country, was involved in a road accident that eventually led to his death in 2005. Yet another example is the death, last year, of Abubakar Rimi, the first civilian governor of Kano State. He is reported to have died waiting for emergency attention at the Mallam Aminu Kano Teaching Hospital in Kano. There are many others but these few examples show how connected we all are in experiencing the negative consequences of our current situation and thereby illustrate that no one is spared. There is no need to discuss whether or not these people failed in their roles. The point is that every point of failure or inefficiency eventually leads to collateral damages that can spread in anydirection.

Most people claim that the fight against corruption is failing due to the lack of enforcement of existing laws. And while that sounds true, I want us to examine that claim a bit further. Lack of enforcement is usually taken to mean that people are not getting punished for doing wrong. While that can explain why more wrong will be done and therefore be the reason for the scale and levels of impunity being displayed today, it does not explain the existence of corruption. People do not start doing the wrong things, in the first instance, just because they may not be caught. There must also be a need to meet.

Based on the foregoing, enforcement may reduce the scale or delay the growth but will not eliminate corruption. In other words, enforcement is a necessary but not a sufficient condition for success in the fight against corruption. And some might say that whatever we get from enforcement is good enough but if we are looking at a drastic reduction in levels of corruption currently seen, we need to look at root causes and we need to look beyond enforcement.

Enforcement is expensive because a lot of resources are needed to effectively police this country. The effectiveness of enforcement also depends heavily on the publicity given to the successes of the relevant agencies. If wrong-doers are getting punished and people do not know, they may still carry on normally believing things are business as usual. More to the point is the fact that the fight against corruption through enforcement is being fought to a standstill not just by corruption itself but also by the inefficiency of our judiciary. Hence, we read of many arrests and suspects but few convictions.

The above instances look at enforcement from the point of view of the government. But even on individual levels things are not much better. We know so many people around us; relations and friends, who get involved in corrupt acts on a daily basis, but we rarely caution them or report them not just for sentimental reasons or because we do not trust the police but also because, at some level, we understand their situation. So, we find it infinitely easier to rail against corruption when it is either perpetrated against us or by someone we do not know. But what is it that we understand? We understand that a lot of Nigerians do not go out wanting to commit crimes but frequently find themselves in situations where the corrupt way becomes the sensible thing to do. Like when you are alone in a deserted road with policemen who insist that N5000 will temporarily clear you of car-smuggling charges. As you see them glance longingly at your purse and at the same time fiddle with their guns you may quickly decide to see their point of view and choose another opportunity to progress the fight againstcorruption.

In their recent report titled "Nigeria's Elections: Reversing the Degeneration?", the International Crisis Group warn that the average politician going into elections today is faced with a dilemma: whether to do the right thing and risk losing to corrupt politicians or to rig the elections themselves knowing that they are unlikely to be caught. That is the same dilemma most Nigerians face on a daily basis with respect to corruption; do things right and suffer or do the corrupt thing and risk getting caught. Knowing that the chances of getting caught are low makes it easier for most people to choose the latter option. But it goes beyond that because in Nigeria today, the corrupt way is frequently the more effective way to get things done and that makes corrupt way a fait accompli.

And that is where the other way of looking at lack of enforcement is useful. Enforcement should not just be seen as punishing people for doing wrong. It should also extend to include rewarding those who do right. So, when people who do right are not rewarded or are maybe even punished for doing right, that is also lack of enforcement. It is my claim that, corruption thrives in societies where people are not provided enough legitimate means of meeting their legitimate needs. When that happens, people who have to meet their needs anyway they can, turn to available, effective ways whether or not such ways are corrupt. I will illustrate with 2 examples; one from the recent Voter Registration Exercise and another from the E-passport Registration Process.

In the early days of the Voter Registration Exercise, in most polling booths where things worked well, people queued up properly, waited their turn, registered and received their cards. Those who did not succeed planned for another day. But as days progressed and the demand for voter's cards exceeded supply, as the threat of government agents increased the importance of owning a card, desperation set in. People were no longer sure that they would obtain cards in the available time. Queuing and being patient became a poor strategy and alternative arrangements were needed... and provided. We eventually heard of locations where people just showed up, paid and were given express service.

The E-passport Registration Process is another example where people find that going the official route of making online payments rarely works. So, a lot of people pay more than the official rates in order to get their passports on time through unofficial means. In these 2 instances there were legitimate needs to meet but the more effective way to meet them was the corrupt way.

I will define corruption as the perversion of official activities for inappropriate private gain. This definition covers both public and private sector and also captures wrong-doing from the perspective of the taker as well as the giver of bribes. While the giver participates, at least, in "the perversion of official activities", the taker adds "inappropriate private gain" to his crime.

However, no matter the culpability of the parties involved, it is obvious that, in most cases, the activities are rooted in meeting legitimate needs. There is nothing therefore peculiar about Nigerians or citizens of other African or Third World countries where corruption is rife, except in the fact that these people live in societies which do not provide enough legitimate avenues for their citizens to meet their needs. This is also proven in the fact that when a lot of these people find themselves in well-run societies with robust processes, they adopt those processes and thrive in them and when they encounter corruption in such societies they do not just sit back and say this is familiar. They usually challenge such corrupt practices using available mechanisms. The corollary is also that when foreigners visit Nigeria they tend to adopt the corrupt approaches irrespective of their own previous experiences. This is readily witnessed by the many international companies who have gone afoul of the US Foreign Corrupt Practices Act due to the activities of their Nigerian subsidiaries.

In the circumstances we have described, where the official way is difficult, insisting, through enforcement, that people do things the tough way simply because it is right is asking people to work against their natural inclinations. Asking people to choose between meeting their needs through legitimate but difficult ways or through corrupt and easy ways is not much of an option. Natural thing is to go for easier. Lack of enforcement only lowers the barrier and makes decision-making easier.

This therefore means that the other crucial aspect to fighting corruption is in making both existing and future legitimate approaches effective. People are not enamoured of the corrupt way (except for the habitual criminals who law enforcement will eventually catch up with) but need effective means of meeting their needs. If the legitimate way is made easy, simple and fast, human nature makes it the obvious choice thereby invalidating the corrupt way. In the Voter Registration Exercise example given above, making the legitimate way effective would mean providing enough resources or extending the exercise over a sufficient period of time to enable smooth registration. For E-passport registration, this means ensuring the online system works while providing other options for people who are not online.

Another approach that complements making legitimate approaches effective is to consider legitimizing currently effective but corrupt approaches, wherever possible. Bearing in mind that the needs are legitimate and it is usually the approach that is corrupt, what this then entails is to review existing practices and identify those useful ones which are not backed up by existing rules and adopt them. Some things that are widely acknowledged as sensible are illegal simply because they have not been accommodated in existing rules. By correcting such, everybody is free to operate and everything is done in the open. This might not strictly apply to both examples given above but we can look at possible scenarios.

In the Voter Registration Exercise, INEC could, in theory, consider setting up a Quick Booth where people with valid reasons could be served separately. What those valid reasons are or how this will be implemented are details that must be worked out. Similarly, I think the Nigerian Immigration may already have an option for quick registration. What I am unsure of is whether it is implemented officially or unofficially. The quick action would then be to make the unofficial official to complement the existing online facility.

Although these two broad strategies can quickly turn the tables on corruption, they are by no means easy to implement.

Firstly, it is difficult to get things right first time. And that is why services are supposed to be tested rigorously or carried out in small pilot projects to ensure robustness before rolling out to the public. It is therefore necessary that process owners remain accountable for ensuring that their processes work effectively.

Secondly, an effective feedback mechanism should accompany these processes. Mechanisms such as Customer Feedback phone lines, suggestion boxes, online forms and toll-free lines are necessary to enable active collaboration with customers. Every time, the system works against people, they will happily give feedback especially when they believe there is genuine interest in improving things.

Thirdly, there is a need to regularly review existing practices with a view to identifying obsolete practices and improving them. This is particularly necessary because the allure of the corrupt way is always in presenting itself as the more effective approach and so those who perpetrate corruption will go out of their way to make the official way ineffective. That is why, for instance, the E-passport online payment system will not work most of the time. If process owners keep their customers in view and work with them, they will identify genuine opportunities for improvement.

Lastly, all of these must be backed up by an effective judiciary which serves as the arbiter for aggrieved parties. An effective judiciary is one which, in my opinion, does not just deliver impartial judgments but delivers them fast enough to make a difference in the issue at hand.

The breadth of corruption cuts across every area of our lives in Nigeria ranging from the judiciary through to vehicle license registration and these strategies are applicable to most. The main point made here is that corruption thrives today not because Nigerians are bad people or just because people are not punished enough but mainly because a lot of systems are not working effectively. There is no doubt that in a lot of the cases we see, the systems are being deliberately under-mined. But the point is that in allowing such ineffective systems to continue and focusing more on people's behaviors, we encourage the spread of corruption. For a lot of people, it is the structure in place that drives their behaviour so to change behaviour, what is required is to change the structures.

I think it worth noting that this discussion seems to leave the desired change in the hands of those who own these systems and therefore seem to currently benefit from corruption but that is not entirely true. Most of the time, corruption works against most of us (and benefits a few in the short term) and that creates a lot of interested passionate parties. If these people are empowered, their passion will drive change through irrespective of the desires of the few who benefit from the status quo. That empowerment resides mainly with the judiciary – another system that desperately needs improvement and which may therefore be the first port of call for any genuine anti-corruption exercise.

Author's Insight

"Like slavery, like apartheid, poverty is not natural.
It is man-made and can be overcome by
the actions of human beings"
— Nelson Mandela.

Isn't this a paradox? Throughout the developing world, governments, companies, organisations and institutions have been trying to eliminate global poverty to no avail, yet the world of economic growth and technological innovation are moving faster than ever. Ending poverty is a big and complex challenge indeed. It has many causes, and there is certainly no silver bullet or single solution, but this complexity is solvable. There are a huge number of smart and talented people all over the developing world, doing amazing work in business, technology and entrepreneurship. There is a need to further leverage their works, skills and collective numbers for increased impact and visibility – this is why I want to join hands and contribute through my book – Digital Transformation: Evolving a digitally enabled Nigerian Public Service.

Growing up in Nigeria, I have come to appreciate first-hand the pains and challenges associated with poverty, often attributable to literacy, ignorance, disease, miss-information or the total lack of it not to mention climate change, health care, good governance and energy access. Digital Transformation will open the eyes of the public especially professionals young and old to the rampant problems in the country and will encourage readers to join in moving into the future of a better and stronger Nigerian that will curb illiteracy, ignorance and disease.

"Digital Transformation: Evolving a Digitally Enabled Nigerian Public Service" was written to bridge this existing knowledge gap and provide clear answers to the reader looking for answers. It is written in a clear but detailed manner, so as to provide a guide towards understanding new technology solution concepts, the specific scenarios where they can be applied especially in the context of developing environments, and the best approaches to be taken towards implementing such solutions while being aware of critical issues impacting execution. It would serve as a long-term solution reference that is relevant at all stages of moving government forward through a digital transformation effort. The final goal is to empower governments of developing countries like Nigeria with the knowledge and tools required to provide sustainable long-term solutions for and to the benefit of their people.

About The Author

Jacobs Edo

Jacobs Edo is the ERP Systems Coordinator at the OPEC Fund for International Development (OFID), Vienna, Austria. He is an enterprise architect, digital transformation manager, trusted advisor to C-Suite, public speaker and author. He is known for his dedication and tireless efforts in ethics, change management and digital transformation. He earned a Bachelor of Engineering degree in Electrical Engineering (Second Class Upper Division) from the University of Nigeria, Nsukka, a diploma in Information Technology from Robert Gordon University, Aberdeen and SPDC, an MSc degree in Telecommunication and Internet Technologies from the Technical University of Applied Sciences, Vienna and a Global Business Transformation Management (GBTM) Master's certification by SAP AG and the University of Applied Sciences and Arts, Northwestern Switzerland's Business Transformation Academy.

As a volunteer Consultant for the Stanford Institute for Innovation in Developing Economies (Stanford SEED), he supports its vision of bringing the power of innovation, entrepreneurship, and leadership to established businesses in sub-Saharan Africa and their leaders on the ground and in their communities.

Meet the Author:
Website: www.jacobsedo.com
Email: me@jacobsedo.com
Twitter: @JacobsEdo

Subscribe to Jacobs Edo's digital transformation reading list Click here for additional reading resources.

Index

www.ingramcontent.com/pod-product-compliance
Lightning Source LLC
Chambersburg PA
CBHW080619060326
40690CB00021B/4748